# C++
# 多线程编程实战

［黑山共和国］Miloš Ljumović 著

姜佑 译

人民邮电出版社

北京

图书在版编目（CIP）数据

C++多线程编程实战 ／（黑）留莫维奇著；姜佑译
． -- 北京：人民邮电出版社，2016.5（2023.4重印）
ISBN 978-7-115-41366-6

Ⅰ．①C… Ⅱ．①留… ②姜… Ⅲ．①C语言—程序设计 Ⅳ．①TP312

中国版本图书馆CIP数据核字(2016)第085851号

## 版 权 声 明

Copyright © Packt Publishing 2014. First published in the English language under the title C++ Multithreading Cookbook.
All Rights Reserved.

本书由英国 Packt Publishing 公司授权人民邮电出版社出版。未经出版者书面许可，对本书的任何部分不得以任何方式或任何手段复制和传播。
版权所有，侵权必究。

- ◆ 著　　[黑山共和国] Miloš Ljumović
  　译　　姜　佑
  　责任编辑　傅道坤
  　责任印制　焦志炜
- ◆ 人民邮电出版社出版发行　北京市丰台区成寿寺路11号
  邮编 100164　电子邮件 315@ptpress.com.cn
  网址 http://www.ptpress.com.cn
  北京七彩京通数码快印有限公司印刷
- ◆ 开本：800×1000　1/16
  印张：20　　　　　　2016年5月第1版
  字数：538 千字　　　2023年4月北京第20次印刷
  著作权合同登记号　图字：01-2015-0749 号

定价：69.80 元
读者服务热线：(010)81055410　印装质量热线：(010)81055316
反盗版热线：(010)81055315

# 内容提要

本书是一本实践为主、通俗易懂的 Windows 多线程编程指导。本书使用 C++本地调用，让读者能快速高效地进行并发编程。

全书共 8 章。第 1 章介绍了 C++编程语言的概念和特性。

第 2~5 章介绍了进程、线程、同步、并发的相关知识。其中，第 2 章介绍进程和线程的基本概念，详细介绍了进程和线程对象。第 3 章讲解线程管理方面的知识，以及进程和线程背后的逻辑，简要介绍了线程同步、同步对象和同步技术。第 4 章重点介绍了消息传递技术、窗口处理器、消息队列和管道通信。第 5 章介绍了线程同步和并发操作，讲解了并行、优先级、分发器对象和调度技术，解释了同步对象（如互斥量、信号量、事件和临界区）。

第 6 章介绍.NET 框架中的线程，概述了 C++/CLI .NET 线程对象。简要介绍了托管方法、.NET 同步要素、.NET 线程安全、基于事件的异步模式和 BackgroundWorker 对象，以及其他主题。

第 7~8 章为水平较高的读者准备了一些高级知识，概述了并发设计和高级线程管理。其中，第 7 章讲解理解并发代码设计，涵盖了诸如性能因素、正确性问题、活跃性问题的特性。第 8 章讲解高级线程管理，重点介绍更高级的线程管理知识。详细介绍了线程池的抽象、定制分发对象，以及死锁的解决方案。

附录涵盖了 MySQL Connector C 和 WinDDK 的具体安装步骤，介绍了如何为驱动程序编译和 OpenMP 编译设置 Visual Studio。另外，还介绍了 DebugView 应用程序的安装步骤，并演示了它的使用步骤。

本书主要面向中高级读者，可作为用 C++进行 Windows 多线程编程的参考读物。本书介绍的同步概念非常基础，因此也可作为对这方面技术感兴趣的读者和开发人员的参考书籍。

# 作者简介

Miloš Ljumović 于 7 月 26 日出生在欧洲黑山共和国的首都波德戈里察，在那里度过了小学和中学的时光，还到音乐学校学习了吉他。随后在黑山大学自然科学和数学学院进修了计算机科学。他对计算机有浓厚兴趣，主修操作系统并获得了硕士学位。2009 年 12 月，Miloš 和他的朋友 Danijel 一起成立了自己的公司，并作为一名程序员和高水平的团队一起致力于提供高技术含量的 IT 解决方案。不久，许多资深的开发者加入了他们，合作开发了许多应用程序和系统软件、Web 应用程序和数据库系统。他的客户不仅包括黑山政府，还涉及一些大型的国有企业。他开发了一个新的金融系统 MeNet 以及一些与图片和其他数字媒体类型相关的视频识别软件。除此之外，他还开发了许多网站和其他网络应用程序。他的客户数量众多，不胜枚举。

Miloš 作为国际顾问在美国一家大型的互联网电子商务贸易和数据采集公司工作了几个月。随后于 2014 年 7 月创立了一家新公司：EXPERT.ITS.ME。除了开发软件，他还为 IT 行业的小型企业提供咨询服务，鼓励并帮助他们在处理好企业管理问题的同时，把企业做大做强。另外，Miloš 还是黑山国家委员会（ICT）成员和门萨俱乐部成员。他热爱编程，擅长 C/C++/C#语言，精通 HTML、PHP、TSQL 等，梦想能开发出自己的操作系统。

在业余时间里，Miloš 喜欢打网球、潜水、狩猎和下象棋。他喜欢和自己的团队进行头脑风暴，想出一些在 IT 领域和计算机科学领域新鲜、时尚的好点子。他紧跟 IT 的发展步伐，不断学习新知识、解决新问题。尤其喜欢通过专门课堂和课程教授计算机科学和数学专业的学生，将他们塑造为合格的程序员，帮助他们发现科学之美。想更多了解他的兴趣爱好和近况，请浏览他的公司网站（http://expert.its.me）或个人网站（http://milos.expert.its.me），也可以通过 milos@expert.its.me 与他联系。

# 致 谢

撰写本书，感慨良多。

把本书献给我的父母 Radosav 和 Slavka，以及我的姐姐 Natalija 和 Dušanka。感谢家人无私地奉献和关爱。特别感谢我的母亲，没有她我不可能成为一名程序员。

非常感谢我美丽的妻子 Lara 的付出，她用耐心和爱鼓励着我，无条件地支持我，告诉我不要放弃。我爱你。

感谢我的好友 Danijel，教我如何成为一个成功的商人，激励我每一天都能成为更好的程序员。

感谢黑山大学的老师们，没有他们我不可能成为现在这样的技术专家。特别感谢 Rajko Ćalasan 不厌其烦地教我编程，Milo Tomašević 教我面向对象编程的专业知识，让我从此爱上 C++。特别感谢我最好的老师 Goran Šuković，他经常指导我，带我游历计算机科学的不同领域，让我每天都充满希望，以积极向上的态度学习新的知识。

# 审稿人简介

**Abhishek Gupta** 是印度班加罗尔的一位年轻的嵌入式软件工程师，开发自动车载信息娱乐系统软件多年。Abhishek 于 2011 年在印度理工学院的卡哈拉格普尔理工学院，完成了视觉信息和嵌入系统专业技术硕士的学习。他对视频处理非常感兴趣，喜欢从事嵌入式多媒体系统工作，擅长用 C 和 Linux 进行编程。

欲详细了解他的信息，请访问 www.abhitak.wordpress.com/about-me。

**Venkateshwaran Loganathan** 是一位杰出的软件开发人员，他工作至今从事过设计、开发和软件产品的测试工作，有 6 年多的工作经验。Venkateshwaran 早在 11 岁时就开始通过 FoxPro 学习计算机编程，从那以后，他学习并掌握了多种计算机语言，如 C、C++、Perl、Python、Node.js 和 Unix shell 脚本。他热衷于开源开发，为各种开源技术做出了贡献。

Venkateshwaran 现就职于高知特科技公司（Cognizant Technology Solutions），作为一名技术助理从事物联网领域的研究和开发工作。目前，他活跃于使用射频识别（RFID）设备发展未来技术的概念。在加入高知特科技之前，他已经在一些大型 IT 公司工作多年，如 Infosys、Virtusa 和 NuVeda。自从作为网络开发人员开始了他的职业生涯，他在网络、在线学习、医疗保健等各个专业领域都有涉猎。由于在工作上的突出表现，公司授予了他许多奖项和荣誉。

Venkateshwaran 在安那大学获得计算机科学与工程的学士学位，目前正在攻读比尔拉技术与科学学院软件系统的硕士学位。除编程以外，他还钻研各种技术和软件技能，为新入职的工程师和在校学生授课。Venkateshwaran 喜欢唱歌和徒步旅行，热衷参与社会服务，喜欢和人打交道。欲详细了解他的情况，请访问网站：http://www.venkateshwaranloganathan.com，并给他发邮件：anandvenkat4@gmail.com。

Venkateshwaran 还写了一本书：《PySide GUI 应用程序开发》（*PySide GUI Application Development*），已由 Packt 出版社出版。

我感慨良多。首先，感谢我的母亲 Anbuselvi 和祖母 Saraswathi，感谢她们对我无私的付出，没有她们的支持和帮助，我不可能达到现在的水平。感谢我的兄弟和朋友们。在困难时期，他们一直都不离不弃地帮助我、祝福我。篇幅有限，无法一一列举所有帮助过我的人，我要在这里向他们表达最诚挚的感谢，感谢你们，我生命中的挚友们。

最重要的是，感谢全能的上帝，让我时刻沐浴在他无尽的祝福中。

**Walt Stoneburner** 是一位经验丰富的软件架构师，有超过 25 年的商业应用程序开发和咨询经验，对质量保证、配置管理和安全都有所涉猎。如果刨根问底，他还会承认自己喜欢统计学和编写文档。

Walt 对编程语言设计、协同应用程序、大数据、知识管理、数据可视化和 ASCII 艺术都很感兴趣。他说自己是壁橱极客。Walt 还评估软件产品和消费性电子产品，画漫画，运营一家针对肖像和艺术的自由摄影工作室（CharismaticMoments.com），写一些幽默的段子，用手表演一些小戏法，喜爱游戏设计。此外，他还是一名业余无线电爱好者。

Walt 有一个名为 Walt-O-Matic 的技术博客：http://www.wwco.com/~wls/blog/，通过 wls@wwco.com 或 Walt.Stoneburner@gmail.com 可以直接与他取得联系。

他还参与了其他书籍的审稿：

- *AntiPatterns and Patterns in Software Configuration Management* （ISBN 978-0-471-32929-9，p. xi）
- *Exploiting Software: How to Break Code* （ISBN 978-0-201-78695-8, p. xxxiii）
- *Ruby on Rails: Web Mashup Projects* （ISBN 978-1-847193-93-3）
- *Building Dynamic Web 2.0 Websites with Ruby on Rails* （ISBN 978-1-847193-41-4）
- *Instant Sinatra Starter* （ISBN 978-1782168218）
- *Learning Selenium Testing Tools with Python* （978-1-78398-350-6）
- *Whittier* （ASIN B00GTD1RBS）
- *Cooter Brown's South Mouth Book of Hillbilly Wisdom* （ISBN 978-1-482340-99-0）

**Dinesh Subedi** 是 Yomari 私营有限责任公司的一位软件开发人员，目前从事数据仓库技术和商业智能开发。他毕业于尼泊尔加德满都工程学院（IOE）Pulchowk 学校并获得了计算机工程学士学位。他在 www.codeincodeblock.com 写了四年的博客，发表了许多与 C++软件开发相关的文章。

感谢我的兄弟 Bharat Subedi 在我审校本书时给予我的帮助。

# 前　言

多线程编程正逐渐成为 IT 行业和开发人员关注的焦点。开发商希望开发出用户友好、界面丰富，而且能并发执行的应用程序。强大的 C++语言和本地 Win32 API 特性为多线程编程提供了良好开端。有了强大的 C++，可以轻松地创建不同类型的应用程序，执行并行，而且还能优化现有的工作。

本书是一本实践为主、通俗易懂的 Windows 多线程编程指导。你将学到如何从多线程方案中受益，增强你的开发能力，构建更好的应用程序。本书不仅讲解了创建并行代码时遇到的问题，而且还帮助读者详细理解同步技术。此外，本书还涵盖了 Windows 进程模式、调度技术和进程间通信方面的内容。

本书从基础开始，介绍了最强大的集成开发环境：微软的 Visual Studio。读者将学会使用 Windows 内核的本地特性和.NET 框架的特性。除此之外，本书还详细讲解了如何解决某些常见的并发问题，让读者学会如何在多线程环境中正确地思考。

通过学习本书，读者将学会如何使用互斥量、信号量、临界区、监视器、事件和管道。本书介绍了 C++应用程序中用到的大部分高效同步方式。本书用大量的程序示例，以最好的方式教会读者用 C++开发并发应用程序。

本书使用 C++本地调用，演示如何利用机器硬件来优化性能。本书最终的目标是传授各种多线程概念，让读者能快速高效地进行并行计算和并发编程。

## 本书涵盖的内容

第 1 章，C++概念和特性，介绍了 C++编程语言和许多特性。本章重点介绍了程序的结构、执行流和 Windows OS 运行时对象。详细介绍了结构化方法和面向对象方法。

第 2 章，进程和线程的概念，详细介绍了进程和线程对象。本章涵盖了进程模式背后的思想和 Windows 进程的实现。除此之外，还介绍了进程间通信和典型的 IPC 问题，并简要介绍了在用户空间和内核中的线程实现。

第 3 章，管理线程，介绍了进程和线程背后的逻辑。本章涵盖了 Windows OS 特性，如协作式多任务和抢占式多任务。本章还详细介绍了线程同步以及同步对象和同步技术。

第 4 章，消息传递，重点介绍了消息传递技术、窗口处理器、消息队列和管道通信。

第 5 章，线程同步和并发操作，介绍了并行、优先级、分发器对象和调度技术。本章还解释了同步对象，如互斥量、信号量、事件和临界区。

第 6 章，.NET 框架中的线程，概述了 C++/CLI .NET 线程对象。本章简要介绍了托管方法、.NET 同步要素、.NET 线程安全、基于事件的异步模式和 BackgroundWorker 对象，以及其他主题。

第 7 章，理解并发代码设计，涵盖了诸如性能因素、正确性问题、活跃性问题等特性。通过本章的学习，用户能从另一个更好的视角理解并发和并行应用程序设计。

第 8 章，高级线程管理，重点介绍更高级的线程管理知识。本章详细介绍了线程池的抽象、定制分发对象，以及死锁的解决方案。最后，本章介绍了一个远程线程的示例，演示高级管理。

附录涵盖了 MySQL Connector C 和 WinDDK 的具体安装步骤。另外，还介绍了如何为驱动程序编译和 OpenMP 编译设置 Visual Studio。另外，还介绍了 DebugView 应用程序的安装步骤，演示了它的使用步骤。

## 本书必备软件

要运行本书中的示例，必须安装下面的软件。

- Visual Studio 2013
    - http://www.visualstudio.com/downloads/download-visual-studio-vs#d-express-windows-8

- Windows 驱动程序套件：WinDDK
    - http://msdn.microsoft.com/en-us/windows/hardware/hh852365.aspx

- MySQL Connector C
    - http://dev.mysql.com/downloads/connector/c/

## 本书的读者对象

本书主要面向中级和高级水平的读者。本书介绍的同步概念非常基础，因此也可作为对这方面技术感兴趣的所有读者和开发人员的参考书籍。最后两章为水平较高的读者准备了一些高级知识,概述了并发设计和高级线程管理等主题。

## 本书的体例

本书通过不同的文本样式以区别不同类型的信息。这里介绍一下这些样式的示例,并解释其含义。

本书出现的代码、数据库表名、文件夹名称、文件名、文件扩展名、路径名、用户输入如下所示:"我们可以通过使用 include 指令包含其他上下文"。

代码块如下所示:

```
class CLock
{
public:
  CLock(TCHAR* szMutexName);
  ~CLock();
private:
  HANDLE hMutex;
};

inline CLock::CLock(TCHAR* szMutexName)
{
    hMutex = CreateMutex(NULL, FALSE, szMutexName);
    WaitForSingleObject(hMutex, INFINITE);
}

inline CLock::~CLock()
{
    ReleaseMutex(hMutex);
    CloseHandle(hMutex);
}
```

通过加粗的方式提醒读者注意代码块中的某些部分:

```
class CLock
{
public:
  CLock(TCHAR* szMutexName);
  ~CLock();
private:
  HANDLE hMutex;
};

inline CLock::CLock(TCHAR* szMutexName)
{
    hMutex = CreateMutex(NULL, FALSE, szMutexName);
    WaitForSingleObject(hMutex, INFINITE);
}
```

```
inline CLock::~CLock()
{
    ReleaseMutex(hMutex);
    CloseHandle(hMutex);
}
```

复制和粘贴本书中的代码时，要特别注意。由于书页篇幅有限，一些代码没法都放在一行，不得不分成多行。我们尝试解决这个问题，但是在某些情况下不太可能。我们强烈建议读者在开始编译示例之前，检查每一个示例，特别是那些带双引号的字符串。由于某些特殊情况，代码被分成多行显示，但要读者要明白这些代码是不能分行的，否则无法通过编译。

在"操作步骤"中，当文中写道"添加第 1 章的现有头文件 CQueue.h"时，我们的意思是：读者要使用 Windows 资源管理器，导航至存放有 CQueue.h 文件的文件夹，并且把该文件以及与之相关的 CList.h 复制到当前项目的工作文件夹中。然后在 Visual Studio 中通过【新增】-【现有项】，把该文件添加至项目中。这样才能正确地编译运行示例代码。

**新术语**和**重要的文字**加粗显示。菜单上的选项以这种形式显示：打开【解决方案资源管理器】，右键单击【头文件】。

 警告或重要的事项。

# 目　　录

## 第 1 章　C++概念和特性简介 ..................................................................... 1
### 1.1　介绍 ............................................................................................................... 1
### 1.2　创建 C++项目 ............................................................................................... 2
### 1.3　程序结构、执行流和运行时对象 ............................................................... 3
### 1.4　结构化编程方法 ........................................................................................... 7
### 1.5　理解面向对象编程方法 ............................................................................... 9
### 1.6　解释继承、重载和覆盖 ............................................................................. 11
### 1.7　理解多态 ..................................................................................................... 15
### 1.8　事件处理器和消息传递接口 ..................................................................... 18
### 1.9　链表、队列和栈示例 ................................................................................. 22

## 第 2 章　进程和线程的概念 ........................................................................ 31
### 2.1　简介 ............................................................................................................. 31
### 2.2　进程和线程 ................................................................................................. 31
### 2.3　解释进程模型 ............................................................................................. 32
### 2.4　进程的实现 ................................................................................................. 36
### 2.5　进程间通信（IPC） ................................................................................... 39
### 2.6　解决典型的 IPC 问题 ................................................................................. 47
### 2.7　线程模型的实现 ......................................................................................... 55
### 2.8　线程的用法 ................................................................................................. 60
### 2.9　在用户空间实现线程 ................................................................................. 66
### 2.10　在内核实现线程 ....................................................................................... 73

## 第3章 管理进程 ............................................................. 79

### 3.1 介绍 ............................................................. 79
### 3.2 进程和线程 ............................................................. 80
### 3.3 协作式和抢占式多任务处理 ............................................................. 83
### 3.4 解释 Windows 线程对象 ............................................................. 84
### 3.5 基本线程管理 ............................................................. 85
### 3.6 实现异步的线程 ............................................................. 92
### 3.7 实现同步的线程 ............................................................. 97
### 3.8 Win32 同步对象和技术 ............................................................. 101
#### 3.8.1 同步对象：互斥量 ............................................................. 102
#### 3.8.2 同步对象：信号量 ............................................................. 103
#### 3.8.3 同步对象：事件 ............................................................. 104
#### 3.8.4 同步对象：临界区 ............................................................. 105

## 第4章 消息传递 ............................................................. 107
### 4.1 介绍 ............................................................. 107
### 4.2 解释消息传递接口 ............................................................. 108
### 4.3 理解消息队列 ............................................................. 112
### 4.4 使用线程消息队列 ............................................................. 118
### 4.5 通过管道对象通信 ............................................................. 122

## 第5章 线程同步和并发操作 ............................................................. 127
### 5.1 介绍 ............................................................. 127
### 5.2 伪并行 ............................................................. 127
### 5.3 理解进程和线程优先级 ............................................................. 128
### 5.4 Windows 分发器对象和调度 ............................................................. 134

5.5 使用互斥量 ........................................................................................ 135
5.6 使用信号量 ........................................................................................ 143
5.7 使用事件 ............................................................................................ 150
5.8 使用临界区 ........................................................................................ 157
5.9 使用管道 ............................................................................................ 164

## 第6章 .NET 框架中的线程

6.1 介绍 .................................................................................................... 177
6.2 托管代码和非托管代码 .................................................................... 177
6.3 如何在.NET 中运行线程 .................................................................. 179
6.4 前台线程和后台线程的区别 ............................................................ 185
6.5 理解.NET 同步要素 .......................................................................... 188
6.6 锁和避免死锁 .................................................................................... 193
6.7 线程安全和.NET 框架的类型 .......................................................... 198
6.8 事件等待句柄的触发 ........................................................................ 200
6.9 基于事件的异步模式 ........................................................................ 204
6.10 BackgoundWorker 类 ...................................................................... 210
6.11 中断、中止和安全取消线程执行 .................................................. 214
6.12 非阻塞同步 ...................................................................................... 222
6.13 Wait 和 Pulse 触发 .......................................................................... 224
6.14 Barrier 类 .......................................................................................... 228

## 第7章 理解并发代码设计

7.1 介绍 .................................................................................................... 235
7.2 如何设计并行应用程序 .................................................................... 235
7.3 理解代码设计中的并行 .................................................................... 240
7.4 转向并行 ............................................................................................ 246

7.5 改进性能因素 ..................................................................................252

## 第8章 高级线程管理 ..................................................................259
8.1 介绍 ..................................................................................259
8.2 使用线程池 ..........................................................................259
8.3 定制线程池分发器 ..................................................................269
8.4 使用远程线程 ........................................................................283

## 附录 A ..............................................................................291
A.1 安装 MySQL Connector/C ........................................................291
A.2 安装 WinDDK-Driver 开发套件 ..................................................294
A.3 设置驱动器编译的 Visual Studio 项目 ..........................................296
A.4 使用 DebugView 应用程序 ......................................................301
A.5 设置 OpenMP 编译的 Visual Studio 项目 ....................................302

# 第 1 章　C++概念和特性简介

**本章介绍以下内容：**
- 创建一个 C++项目
- 程序结构、执行流、运行时对象
- 结构编程方法
- 理解面向对象编程方法
- 解释继承、重载和覆盖
- 理解多态
- 事件处理器和消息传递接口
- 链表、队列、栈示例

## 1.1 介绍

系统所执行的进程或抽象是所有操作系统的核心概念。现在绝大多数的操作系统在同一时间内都可以进行多项操作。例如，计算机在用户编辑 Word 文档时，还可以打印该文档、从硬盘缓冲区读数据、播放音乐等。在多任务操作系统中，中央处理单元（CPU）在程序中快速切换，执行每个程序只需几毫秒。

从严格意义上来说，对于单处理器系统，处理器在一个单元时间内只能执行一个进程。操作系统以极快的速度切换多个进程，营造了一个多进程同时运行的假象。与多处理器系统中硬件支持的真正并行相比，单处理器系统的这种并行叫伪并行（*pseudoparallelism*）更合适。

多线程（*multithreading*）是现代操作系统中非常重要的概念。多线程即允许执行多个线程，对完成并行任务和提升用户体验非常重要。

在传统的操作系统中，每个进程都有自己的地址空间和一个执行线程，该线程通常叫主线程（*primary thread*）。一般而言，运行在同一个进程中的多个线程具有相同的地址空间（即进程的地址空间），在准并行上下文中，这些线程就像是多个单独运行的进程，只不过它们的地址空间相同。

> 伪并行是操作系统在单处理器环境下的特性。准并行地址空间概念是 Windows 操作系统的特性。在多处理器系统中，Windows 为每个进程提供了一个虚拟地址空间，比真正的物理地址空间大得多，因此叫做准并行上下文。

线程（*thread*）是操作系统中的一个重要概念。线程对象包含一个程序计数器（负责处理在下一次线程获取处理器时间时要执行什么指令）、一组寄存器（储存线程正在操控的变量当前值）、一个栈（储存与函数调用和参数相关的数据），等等。虽然线程执行在进程的上下文中，但是它们的区别很大。进程非常贪婪，想占用所有的资源；而线程比较"友好"，它们彼此合作、交流，而且共享资源（如处理器时间、内存和变量等）。

## 1.2　创建 C++项目

本书所有的程序示例均在 Visal Studio IDE 中运行。下面，针对 Visual Studio 介绍如何正确地设置 IDE，并指出一些影响多线程应用程序的具体设置。

### 准备就绪

确定安装并运行了 Visual Studio（VS）。

### 操作步骤

运行 Visual Studio，在【开始】界面选择【新建项目】，会弹出一个有多个选项的窗口。在左边【模板】下面，选择【C++】，展开 C++节点，有【CLR】、【常规】、【测试】、【Win32】等选项。然后，执行以下步骤。

1. 选择 Win32。在中间栏有两个选项：【Win32 控制台应用程序】和【Win32 项目】。
   目前，我们使用【Win32 控制台应用程序】。【Win32 项目】用于有**图形用户接口**（GUI）的应用程序，而不是控制台程序。如果使用控制台，要在项目属性中设置其他选项。

2. 选择【Win32 控制台应用程序】，并在窗口下方的【名称】右边为项目命名。我们把第 1 个 Win32 控制台应用程序项目命名为 TestProject。在【位置】右边选择储存该项目文件的文件夹。VS 将帮你创建一个文件夹，把用户刚才在【位置】输入的文件夹作为将来创建项目的默认文件夹。
   现在，读者应该看到 Win32 应用程序向导窗口。可以直接单击右下方的【完成】，这样 VS 会自动创建所有需要的文件。或者，选择【下一步】，然后在附加选项中勾选【空项目】。如果这样做，就要自己创建源文件和头文件，VS 不会自动生成所需的文件。

3. 如果在上一步骤的 Win32 应用程序向导窗口中直接选择【完成】，在【解决方案资源管理器】中就可以看到 stdafx.h 和 targetver.h 头文件，以及 stdafx.cpp 和 TestProject.cpp 源文件。stdafx.h 和 stdafx.cpp 文件是预处理头文件的一部分，用于智能感应引擎。该引擎使用翻译单元（*Translation Unit*，TU）模型模仿命令行编译器，用于智能感应。典型的翻译单元由一个源文件和包含在源文件中的多个头文件组成。当然，其中还引用了其他头文件，所以也包含这些被引用的头文件。智能感应引擎从一个特殊的子串开始，给用户提供信息（如，特定类型是什么、函数和重载函数的原型是什么、在当前作用域中变量是否可用等）。欲了解更多相关内容，请查阅 MSDN

参考资料（http://msdn.microsoft.com）。

4. TestProject.cpp 文件出现在中间的窗口，这就是编写代码的地方。以后，我们会在更复杂的项目中创建和使用更多的文件，现在先暂时介绍这么多。

### 示例分析

每个程序都必须有自己的主例程，即 main。当运行程序时，操作系统从调用 main 开始。这是执行 C++ 程序的起点。如果编写的代码遵循 Unicode 编程模型，就可以使用 main 的宽字符版本 wmain。当然，也可以使用定义在 TCHAR.h 中的_tmain。如果定义了_UNICODE，_tmain 函数相当于 wmain 函数；如果没有定义_UNICODE，_tmain 函数相当于 main 函数。

在 TestProject 窗口上方，有各种各样的按钮和选项。其中有一个包含 Win32 可选项的下拉菜单，这个选项叫做【解决方案平台】。如果要创建 32 位可执行文件，就不用改动。如果要创建 64 位可执行文件，先展开下拉菜单，选择【配置管理器】，找到【活动解决方案平台】，选择【x64】选项。点击【确定】，然后关闭【配置管理器】窗口。

在创建 64 位可执行文件时，最重要的是更改项目属性中的设置。按下 **Alt+F7**，或者右键单击【解决方案资源管理器】中的 TestProject 项目，选择【属性】，弹出 TestProject 属性页窗口。在【配置属性】的【C/C++】的下拉菜单中选择【预处理器】。在【预处理器定义】中，把 WIN32 改成_WIN64 才能创建 64 位可执行文件。其他设置暂不更改。

无论创建 32 位还是 64 位的代码，都要正确设置代码生成。创建 C++项目时，可以选择该应用程序是否依赖用户 PC 上 C++运行时所需的**动态链接库**（DLL）。如果创建的应用程序不仅在本机上运行，还要在其他 PC 上运行，就要考虑这一点。用 VS 在本机开发应用程序，所需的 C++运行时库已经安装，不会有任何问题。但是，在其他未安装 C++运行时库的 PC 上运行这种应用程序，就有可能出问题。如果确认不依赖 DLL，则需把【运行时库】选项改为【多线程调试（/MTd）】的调试模式，或改为【多线程（/MT）】发布模式。调试模式或发布模式在【解决方案配置】的下拉菜单中可任意切换。

对于本书的程序示例，其他选项都不需要改动，因为 32 位和 64 位的机器都能运行 32 位可执行文件。运行时库作为 C++软件包框架已经安装在 PC 中了，使用默认设置即可，应用程序在这样的 PC 中运行没有问题。

## 1.3 程序结构、执行流和运行时对象

编程范式是计算机编程的基本样式，主要有 4 种范式：命令式、声明式、函数式（或结构式）、面向对象。C++是当今最流行的面向对象编程语言之一，集功能性、灵活性、实用性于一体。和 C 一样，程序员能很快地适应它。C++成功的关键在于，程序员可以根据实际需要做相应地调整。

但是，C++学起来并不轻松。有时，你会认为这是一门高深莫测、难以捉摸的语言，一门永远学不完也无法完全理解和掌握的语言。别担心，学习一门语言并不是要掌握它的所有细枝末节，关键要学会如何正确地用语言特性解决特定的问题。实践是最好的老师，根据具体情况尽可能多地使用相应的特性，有助于加深理解。

在给出示例前，我们先介绍一下查尔斯·西蒙尼的匈牙利表示法。他在 1977 年的博士论文中，使用元编程（*Meta-Programming*）（一种软件生产方法）在程序设计中制定了标准的表示法。文中规定类型或变量的第 1 个字母表示数据类型。例如，如果要给一个类命名，Test 数据类型应该是 CTest。第 1 个字母 C 表示 Test 是一个类。这个方法很不错，因为不熟悉 Test 数据类型的程序员会马上明白 Test 是一个类名。基本数据类型也可以这样处理，以 int 和 double 为例，iCount 表示一个 int 类型的变量 Count，而 dValues 表示一个 double 类型的变量 Value。有了这些前缀，即使不熟悉代码也很容易识别它们的类型，提高了代码的可读性。

## 准备就绪

确定安装并运行了 Visual Studio（VS）。

## 操作步骤

根据以下步骤创建我们的第 1 个程序示例。

1. 创建一个的默认的 C++控制台应用程序[1]，命名为 TestDemo。
2. 打开 TestDemo.cpp。
3. 输入下面的代码：

   ```
   #include "stdafx.h"
   #include <iostream>

   using namespace std;

   int _tmain(int argc, _TCHAR* argv[])
   {
      cout << "Hello world" << endl;
      return 0;
   }
   ```

## 示例分析

不同的编程技术使得 C++程序的结构多种多样。绝大多程序都必须有#include 或预处理指令。

#inlude <iostream>告诉编译器包含 iostream.h 头文件，该头文件中有许多函数原型。这也意味着函数实现以及相关的库都要放进可执行文件中。因此，如果要使用某个 API 或函数，就要包含相应的头文

---

[1] 译注：即弹出向导窗口后直接点【完成】。

件，必要时还必须添加包含 API 或函数实现的输入库。另外，要注意<header>和"header"的区别。前者（< >）表示从解决方案配置的项目路径开始搜索，而后者（" "）表示从与 C++项目相关的当前文件夹开始搜索。

using 命令指示编译器要使用 std 名称空间。名称空间中包含对象声明和函数实现，有非常重要的用途。在包含第三方库时，名称空间能最大程度地减少两个不同软件包中同名函数的歧义。

我们需要实现一个程序的入口点：main 函数。前面提到过，ANSI 签名用 main，Unicode 签名用 wmain，编译器根据项目属性页的预处理器定义确定签名用_tmain。对于控制台应用程序，main 函数有以下 4 种不同的原型：

- int _tmain(int argc, TCHAR* argv[])
- void _tmain(int argc, TCHAR* argv[])
- int _tmain(void)
- void _tmain(void)

第 1 种原型有两个参数：argc 和 argv。第 1 个参数 argc（即，参数计数）表示第 2 个参数 argv（即，参数值）中的参数个数。形参 argv 是一个字符串数组，其中的每个字符串都代表一个命令行参数。argv 中的第 1 个字符串一定是当前程序的名称。第 2 种原型和第 1 种原型的参数类型、参数个数相同，但是返回类型不同。这说明 main 函数可能返回值，也可能不返回值。该值将被返回给操作系统。第 3 种原型没有参数，并返回一个整型值。第 4 种原型既没有参数也没有返回类型。看来，用第 1 种原型作为练习很不错。

函数体中的第 1 条语句使用了 cout 对象。cout 是 C++中标准输出流的名称。整条语句的意思是：把一系列字符（该例中是 Hello world 字符序列）插入标准输出流（通常对应的是屏幕）。

cout 对象声明在 std 名称空间的 iostream 标准文件中。因此，要使用该对象必须包含相应的头文件，并且在_tmain 函数前面先声明其所属的名称空间。

在我们使用的原型中（int _tmain(int, _TCHAR*)），_tmain 返回一个整数。因此，必须在 return 关键字后面指定相应的 int 类型值，本例中是 0。向操作系统返回值时，0 通常表示执行成功。但是，具体的值由操作系统决定。

这个小程序非常简单。我们以此为例，解释 main 例程作为每个 C++程序入口点的基本结构和用法。

单线程程序按顺序逐行执行。因此，如果把所有的代码都写成一个线程，这样的程序对用户并不友好。

如图 1.1 所示，应用程序要等用户输入数据后，才能重新获得控制权继续执行。为此，可以创建并发线程来处理用户的输入。这样，应用程序随时都能响应，不会在等待用户输入时毫无反应了。线程处理完自己的任务后，可以给应用程序发信号，告诉程序用户已完成相应操作。

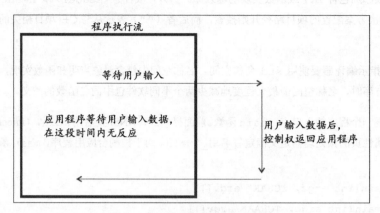

图 1.1　单线程程序按顺序逐行执行

## 更多讨论

每次我们要在主执行流中单独执行一个操作，都必须考虑使用一个单独的线程。最简单的例子是，实现一边计算一边在进度条上反映计算的进度。想在同一个线程中处理计算和更新进度条，可能行不通。因为如果一个线程既要进行计算又要更新 UI，就不能充分地与操作系统绘画交互。因此，一般情况下我们总是把 UI 线程与其他工作线程分开。

来看下面的例子。假设我们创建了一个用于计算的函数（如，计算指定角度的正弦值或余弦值），我们要同步显示计算过程的进度：

```
void CalculateSomething(int iCount)
{
    int iCounter = 0;
    while (iCounter++ < iCount)
    {
        // 计算部分
        // 更新进度条部分
    }
}
```

由于 while 循环的每次迭代都忙于依次执行语句，操作系统没有所需的时间逐步更新用户接口（该例中，用户接口指进度条），因此用户见到的可能是空的进度条。待该函数返回后，才会出现已经完全被填满的进度条。出现这种情况的原因是在主线程中创建了进度条。我们应该单独用一个线程来执行 CalculateSomething 函数，然后在每次迭代中给主线程发信号来逐步更新进度条。前面提到过，线程在 CPU 中以极快的速度切换，在我们看来进度条的更新与计算进度同步进行。

总而言之，每次处理并行任务时，如果要等待用户输入或依赖外部（如，远程服务器的响应），就应该为类似的操作单独创建一个线程，这样我们的程序才不会挂起无响应。

我们在后面的示例分析中会讨论静态库和动态库，现在先简要地介绍一下。静态库（*.lib）通常是指一些已编译过的代码，放置在单独的文件中供将来使用。在项目中添加静态库就可以使用它的一些特性了。如前所示，代码中的#include <iostream>指示编译器包含一个静态库，该库包含了一些输入输出流函数的实现。在实际运行程序之前，静态库在编译时被放入可执行文件。动态库（*.dll）与静态库类似，不同的是它不在编译时放入，而在开始执行程序后才进行链接，也就是在运行时链接。如果许多程序都要用到某些函数，动态库就非常有用。有了动态库，就不必在每个程序中都包含这些函数，只需在运行程序时链接一个动态库即可。User32.dll 是一个很好的例子，Windows 操作系统把大部分 GUI 函数都放在该库中。因此，如果创建的两个程序都有窗口（GUI 窗体），不必在两个程序中都包含 CreateWindows，只需在运行时链接 User32.dll，就可以使用 CreateWindows API 了。

## 1.4 结构化编程方法

前面提到过，有 4 种编程范式。用 C++编写程序时，可以使用结构化编程范式和面向对象编程范式。虽然 C++是面向对象语言，但是也能用结构化编程方法来编写程序。通常，一个程序中有一个或多个函数，因为每个程序必须有一个主例程（main）。对大型程序而言，如果把所有代码都放进 main 函数中，会导致代码的可读性非常差。较好的做法是把程序中的代码分成多个处理单元，即函数。接下来，我们用一个计算两个复数之和的程序来说明。

### 准备就绪

确定安装并运行了 Visual Studio。

### 操作步骤

执行下面的步骤。

1. 创建一个新的默认 C++控制台应用程序，命名为 ComplexTest。

2. 打开 ComplexTest.cpp 文件，并输入下面的代码：

    ```
    #include "stdafx.h"
    #include <iostream>

    using namespace std;

    void ComplexAdd(double dReal1, double dImg1, double dReal2,
            double dImg2, double& dReal, double& dImg)
    {
        dReal = dReal1 + dReal2;
        dImg = dImg1 + dImg2;
    }
    ```

```cpp
double Rand(double dMin, double dMax)
{
    double dVal = (double)rand() / RAND_MAX;
    return dMin + dVal * (dMax - dMin);
}

int _tmain(int argc, TCHAR* argv[])
{
    double dReal1 = Rand(-10, 10);
    double dImg1 = Rand(-10, 10);
    double dReal2 = Rand(-10, 10);
    double dImg2 = Rand(-10, 10);
    double dReal = 0;
    double dImg = 0;
    ComplexAdd(dReal1, dImg1, dReal2, dImg2, dReal, dImg);
    cout << dReal << "+" << dImg << "i" << endl;
    return 0;
}
```

## 示例分析

我们创建了 ComplexAdd 函数，它有 6 个参数（或者 3 个复数）。前两个参数分别是第 1 个复数的实部和虚部；第 3 和第 4 个参数分别是第 2 个复数的实部和虚部；第 5 和第 6 个参数分别是第 3 个复数的实部和虚部，该复数是前两个复数之和。注意，第 5 和第 6 个参数按引用传递。另外，还创建了 Rand 函数，返回一个 dMin 和 dMax 范围内的随机实数。

虽然上面的代码能完成任务，但是读者是否觉得其可读性很差？ComplexAdd 函数的参数太多，main 函数中又有 6 个变量。看来，这种处理方法并不"友好"。我们稍微改进一下，如下所示：

```cpp
#include "stdafx.h"
#include <iostream>

using namespace std;

struct SComplex
{
    double dReal;
    double dImg;
};

SComplex ComplexAdd(SComplex c1, SComplex c2)
{
    SComplex c;
    c.dReal = c1.dReal + c2.dReal;
    c.dImg = c1.dImg + c2.dImg;
    return c;
}

double Rand(double dMin, double dMax)
{
    double dVal = (double)rand() / RAND_MAX;
    return dMin + dVal * (dMax - dMin);
}
```

```cpp
int _tmain(int argc, TCHAR* argv[])
{
    SComplex c1;
    c1.dReal = Rand(-10, 10);
    c1.dImg = Rand(-10, 10);

    SComplex c2;
    c2.dReal = Rand(-10, 10);
    c2.dImg = Rand(-10, 10);

    SComplex c = ComplexAdd(c1, c2);

    cout << c.dReal << "+" << c.dImg << "i" << endl;
    return 0;
}
```

### 更多讨论

这次，我们创建了一个新类型（在该例中是结构）SComplex 来表示复数，提高了代码的可读性，而且比之前的例子更有意义。因此，可以在代码中通过创建对象来执行抽象的任务，这样做提高了程序本身的逻辑性和可读性，以这种方式编程更容易。

## 1.5　理解面向对象编程方法

面向对象编程（OOP）是为真实世界创建软件模型的全新方法，它以一种独特的方式设计程序。OOP 有几个核心概念，如类、对象、继承和多态。

### 准备就绪

确定安装并运行了 Visual Studio。

### 操作步骤

我们用 OOP 的方法修改一下前面的示例，请执行下面的步骤。

1. 创建一个新的默认控制台应用程序，命名为 ComplexTestOO。
2. 打开 ComplexTestOO.cpp 文件，输入下面的代码：

```cpp
#include "stdafx.h"
#include <iostream>

using namespace std;

double Rand(double dMin, double dMax)
{
    double dVal = (double)rand() / RAND_MAX;
    return dMin + dVal * (dMax - dMin);
}
```

```cpp
class CComplex
{
public:
    CComplex()
    {
        dReal = Rand(-10, 10);
        dImg = Rand(-10, 10);
    }
    CComplex(double dReal, double dImg)
    {
        this->dReal = dReal;
        this->dImg = dImg;
    }
    friend CComplex operator+ (const CComplex& c1, const CComplex& c2);
    friend ostream& operator<< (ostream& os, const CComplex& c);
private:
    double dReal;
    double dImg;
};

CComplex operator+ (const CComplex& c1, const CComplex& c2)
{
    CComplex c(c1.dReal + c2.dReal, c1.dImg + c2.dImg);
    return c;
}

ostream& operator<< (ostream& os, const CComplex& c)
{
    return os << c.dReal << "+" << c.dImg << "i";
}

int _tmain(int argc, TCHAR* argv[])
{
    CComplex c1;
    CComplex c2(-2.3, 0.9);
    CComplex c = c1 + c2;
    cout << c << endl;

    return 0;
}
```

## 示例分析

查看该例的 main 函数会发现，这和做整数加法的 main 函数差不多。复数并不是一种基本类型，程序员的工作就是添加一个类，并适当设置处理该类的方法。然后，就能像整数那样使用复数了。

在该例中，我们定义了一个新类型 CComple 类。这个自定义的类有自己的特征[2]和方法，当然还可以有能访问其私有成员的友元方法。我们没有改变 Rand 函数，只是让这个表示复数的类尽可能地像复数的抽象。我们创建了 dReal 和 dImg 特征分别表示复数的实部和虚部；创建了 operator+方法，让+（加）对编译器而言有新的含义。改进后的代码可读性提高了，更容易理解，更方便使用。我们还创建了两个构造函

---

[2] 译者注：为了区分 attribute 和 property，本书把 attribute 译为"特征"，property 译为"属性"。

数：一个是默认构造函数（使用-10~10 的随机实数），一个是让用户直接设置复数实部和虚部的构造函数。

如果需要重载某个函数（稍后再详细介绍），有两种方案。第 1 种方案是在类中设置方法，该方法将改变主调对象的状态（或值）。第 2 种方案是把类和方法分开，即把方法放在类的外面，但是该方法必须声明为 friend 才能访问对象中的私有和保护成员。

## 更多讨论

下面的代码告诉编译器，以两个复数为参数的 operator+函数被定义在别处，而且该函数要访问 CComplex 类的私有和保护成员：

```
friend CComplex operator+(const CComplex& c1, const CComplex& c2);
```

这样做尽可能地简化了 main 函数。把用户自定义类型当作基本类型来使用的好处是，即使原本不熟悉该程序的程序员，也能很快地理解代码所要表达的意思。

# 1.6 解释继承、重载和覆盖

继承是 OOP 中非常重要的特性。继承至少关系到两个类（或更多类）：如果 B 类是某一种 A 类，那么 B 类的对象就拥有与 A 类对象相同的属性。除此之外，B 类也可以实现新的方法和属性，以代替 A 类相应的方法和属性。

## 准备就绪

确定安装并运行了 Visual Studio。

## 操作步骤

现在，执行以下步骤来修改前面的示例。

1. 创建一个默认控制台应用程序，命名为 InheritanceTest。

2. 打开 InheritanceTest.cpp 文件，输入下面的代码：

```
#include "stdafx.h"
#include <iostream>

using namespace std;

class CPerson
{
public:
    CPerson(int iAge, char* sName)
    {
```

```cpp
            this->iAge = iAge;
            strcpy_s(this->sName, 32, sName);
        }
        virtual char* WhoAmI()
        {
            return "I am a person";
        }
    private:
        int iAge;
        char sName[32];
    };

    class CWorker : public CPerson
    {
    public:
        CWorker(int iAge, char* sName, char* sEmploymentStatus)
            : CPerson(iAge, sName)
        {
            strcpy_s(this->sEmploymentStatus, 32, sEmploymentStatus);
        }
        virtual char* WhoAmI()
        {
            return "I am a worker";
        }
    private:
        char sEmploymentStatus[32];
    };

    class CStudent : public CPerson
    {
    public:
        CStudent(int iAge, char* sName, char* sStudentIdentityCard)
            : CPerson(iAge, sName)
        {
            strcpy_s(this->sStudentIdentityCard, 32, sStudentIdentityCard);
        }
        virtual char* WhoAmI()
        {
            return "I am a student";
        }
    private:
        char sStudentIdentityCard[32];
    };

    int _tmain(int argc, TCHAR* argv[])
    {
        CPerson cPerson(10, "John");
        cout << cPerson.WhoAmI() << endl;

        CWorker cWorker(35, "Mary", "On wacation");
        cout << cWorker.WhoAmI() << endl;

        CStudent cStudent(22, "Sandra", "Phisician");
        cout << cStudent.WhoAmI() << endl;

        return 0;
    }
```

## 示例分析

我们创建了一个新的数据类型 CPerson 来表示人。该类型用 iAge 和 sName 作为特征来描述一个人。如果还需要其他数据类型来表示工人或学生，就可以用 OOP 提供的一个很好的机制——继承来完成。工人首先是人，然后还有一些其他特征，我们用下面的代码把 CPerson 扩展为 CWorker：

```
class CWorker : public CPerson
```

也就是说，CWorker 从 CPerson 类继承而来。CWorker 类不仅具有基类 CPerson 的所有特征和方法，还有一个对工人而言非常重要的特征 sEmploymentStatus。接下来，我们还要创建一个学生数据类型。除了年龄和名字，学生也还具有其他特征。同理，我们用下面的代码把 CPerson 扩展为 CStudent：

```
class CStudent : public CPerson
```

声明一个对象时，要调用它的构造函数。这里要注意的是：声明一个派生类的对象时，先调用基类的构造函数，后调用派生类的构造函数。如下代码所示：

```
CWorker( int iAge, char* sName, char* sEmploymentStatus )
  : CPerson( iAge, sName )
{
    strcpy_s( this->sEmploymentStatus, 32, sEmploymentStatus );
}
```

注意看 CWorker 构造函数的原型，其形参列表后面有一个：（冒号），后面调用的是基类的构造函数，如下代码所示。在创建 CPerson 时，需要两个参数 iAge 和 sName：

```
CPerson(iAge, sName)
```

调用析构函数的顺序要反过来，即先调用派生类的析构函数，后调用基类的析构函数。

一图胜千言，CPerson、CWorker 和 CStudent 类对象分别如图 1.2 所示。

可以针对用户自定义的类型来定义运算符的含义，如前面例子中的 CComplex。这样做非常好，当 c、c1 和 c2 是复数时，c = c1 + c2 比 c = ComplexAdd(c1, c2) 更直观更容易理解。

要让编译器能处理用户自定义的类型，就必须实现运算符函数或重载相应的函数。假设，有两个矩阵 m1、m2 和一个矩阵表达式 m = m1 + m2。编译器知道如何处理基本类型（如，把两个整数相加），但如果事先没有定义 CMatrix operator+(const CMatrix& m1, const CMtrix& m2)函数，编译器就不知道如何计算矩阵加法。

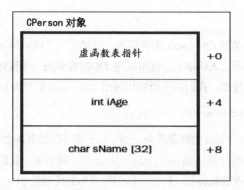
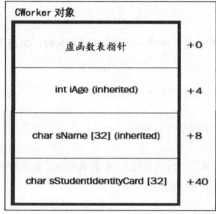
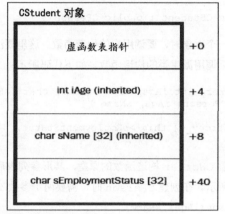

图 1.2 CPerson、CWorker 和 CStudent 类对象

覆盖（override）方法也是一种特性，允许派生类在基类已经实现某方法的前提下提供自己的特定实现。如前面例子中的 WhoAmI 方法所示，其输出如下：

```
I am a person
I am a worker
I am a student
```

虽然每个类中的方法名相同，但它们却是不同的方法，有不同的功能。我们可以说，CPerson 的派生类覆盖了 WhoAmI 方法。

## 更多讨论

覆盖是 OPP 和 C++ 的优异特性，不过多态更胜一筹。我们继续往下看。

## 1.7 理解多态

利用多态（*Polymorphism*）特性，可以通过基类的指针或引用访问派生类的对象，执行派生类中实现的操作。

### 准备就绪

确定安装并运行了 Visual Studio。

### 操作步骤

执行下面的步骤。

1. 创建一个新的默认控制台应用程序，名为 `PolymorphismTest`。
2. 打开 `PolymorphismTest.cpp` 文件，并输入下面的代码：

```cpp
#include "stdafx.h"
#include <iostream>

#define M_PI 3.14159265358979323846

using namespace std;

class CFigure
{
public:
    virtual char* FigureType() = 0;
    virtual double Circumference() = 0;
    virtual double Area() = 0;
    virtual ~CFigure(){ }
};

class CTriangle : public CFigure
{
public:
    CTriangle()
    {
        a = b = c = 0;
    }
    CTriangle(double a, double b, double c) : a(a), b(b), c(c){ }
    virtual char* FigureType()
    {
        return "Triangle";
    }
    virtual double Circumference()
    {
        return a + b + c;
    }
    virtual double Area()
    {
        double S = Circumference() / 2;
        return sqrt(S * (S - a) * (S - b) * (S - c));
```

```cpp
    }
private:
    double a, b, c;
};

class CSquare : public CFigure
{
public:
    CSquare()
    {
        a = b = 0;
    }
    CSquare(double a, double b) : a(a), b(b) { }
    virtual char* FigureType()
    {
        return "Square";
    }
    virtual double Circumference()
    {
        return 2 * a + 2 * b;
    }
    virtual double Area()
    {
        return a * b;
    }
private:
    double a, b;
};

class CCircle : public CFigure
{
public:
    CCircle()
    {
        r = 0;
    }
    CCircle(double r) : r(r) { }
    virtual char* FigureType()
    {
        return "Circle";
    }
    virtual double Circumference()
    {
        return 2 * r * M_PI;
    }
    virtual double Area()
    {
        return r * r * M_PI;
    }
private:
    double r;
};

int _tmain(int argc, _TCHAR* argv[])
{
    CFigure* figures[3];

    figures[0] = new CTriangle(2.1, 3.2, 4.3);
    figures[1] = new CSquare(5.4, 6.5);
```

```
        figures[2] = new CCircle(8.8);
        for (int i = 0; i < 3; i++)
        {
            cout << "Figure type:\t" << figures[i]->FigureType()
                << "\nCircumference:\t" << figures[i]->Circumference()
                << "\nArea:\t\t" << figures[i]->Area()
                << endl << endl;
        }
        return 0;
    }
```

## 示例分析

首先，创建了一个新类型 CFigure。我们想创建一些具体的图形（如，三角形、正方形或者圆），以及计算这些图形周长和面积的方法。但是，我们并不知道具体的图形是什么类型，所以无法用方法直接计算图形的这些特性。这就是要把 CFigure 类创建为抽象类的原因。抽象类是至少声明了一个虚方法的类，该虚方法没有实现，且其原型后面有= 0。以这种方式声明的函数叫做纯虚函数。抽象类不能有对象，但是可以有继承类。因此可以实例化抽象类的指针和引用，然后从 CFigure 类派生出 CTriangle、CSquare 和 CCircle 类，分别表示三角形、正方形和圆形。我们要实例化这些对象的类型，所以在这些派生类中，实现了 FigureType 方法、Circumference 方法和 Area 方法。

虽然这 3 个类中的方法名都相同，但是它们的实现不同，这与覆盖类似但含义不同。如何理解？在本例的 main 函数中，声明了一个数组，内含 3 个 CFigure 类型的指针。作为指向基类的指针或引用，它们一定可以指向该基类的任何派生类。因此，可以创建一个 CTriangle 类型的对象，并设置 CFigure 类型的指针指向它，如下代码所示：

```
figures[ 0 ] = new CTriangle( 2.1, 3.2, 4.3 );
```

同理，用下面的代码可以设置其他图形：

```
figures[ 1 ] = new CSquare( 5.4, 6.5 );
figures[ 2 ] = new CCircle( 8.8 );
```

现在，考虑下面的代码：

```
for ( int i = 0; i < 3; i++ )
{
    cout << "Figure type:\t" << figures[ i ]->FigureType( ) << "\nCircumference:\t"
        << figures[ i ]->Circumference( ) << "\nArea:\t\t" << figures[ i ]->Area( )
        << endl<<endl;
}
```

编译器将使用 C++的动态绑定（*dynamic binding*）特性，确定图形指针具体指向哪个类型的对象，调用合适的虚方法。只有把方法声明为虚方法，且通过指针或引用访问才能使用动态绑定。

现在回头看看 1.6 节的例子。我们分别声明了 CPerson、CWorker 和 CStudent 类型的对象，这是 3 个不同的类型。我们可以通过某个对象调用 WhoAmI 方法，如下代码所示：

```
cPerson.WhoAmI()
```

编译器在编译时就知道 cPerson 对象是 CPerson 类型,也知道 WhoAmI()是该类型的方法。然而在本节的图形例子中,编译器在编译时并不知道图形指针将要指向哪个类型的对象,要等到运行时才知道。因此,这个过程叫动态绑定。

## 1.8 事件处理器和消息传递接口

许多程序都要响应一些事件,例如,当用户按下按键或输入一些文本时。事件处理或程序能响应用户的动作是一种非常重要的机制。如果要在用户按下按键时处理这个事件,就要创建某种监听器,监听按键事件(即,按下的动作)。

事件处理器是操作系统调用的一个函数,每次都发送某种类型的消息。例如,在按下按键时发送"已按下",在文本输入时发送"接收到一个字符"。

事件处理器非常重要。计时器是经过某段时间后触发的事件。当用户按下键盘上的一个按键,操作系统就引发"按下按键"事件,等等。

对我们而言,窗口的事件处理器至关重要。大多数应用程序都有窗口或窗体。每个窗口都要有自己的事件处理器,一旦在窗口中发生事件都要调用事件处理器。例如,如果创建一个带多个按钮和文本框的窗口,则必须有一个与该窗口相关的窗口过程来处理这些事件。

Windows 操作系统以窗口过程的形式提供了这样一种机制,通常命名为 WndProc(也可以叫其他名称)。每次指定窗口发生事件时,操作系统就会调用该过程。在下面的例子中,我们将创建第 1 个 Windows 应用程序(即创建一个窗口),并解释窗口过程的用法。

### 准备就绪

确定安装并运行了 Visual Studio。

### 操作步骤

执行下面的步骤。

1. 创建一个新的 C++ Win32 项目,命名为 GUIProject,单击右下方的【确定】。在弹出的向导窗口中单击【下一步】,在附加选项中勾选【空项目】,然后单击【完成】。现在,在【解决方案资源管理器】中右键单击【源文件】,选择【添加】,然后左键单击【新建项】。在弹出的窗口中选择【C++文件(.cpp)】,命名为 main。然后,单击窗口右下方的【添加】。

2. 现在创建代码。首先,添加所需的头文件:#include <windows.h>

大多数 API 都需要 windows.h 头文件才能处理一些视觉特性,如窗口、控件、枚举和样式。在创建一个应用程序入口点之前，必须先声明一个窗口过程的原型才能在窗口结构中使用它，如下代码所示：

```
LRESULT CALLBACK WndProc(HWND hWnd, UINT uMsg, WPARAM wParam, LPARAM lParam);
```

我们稍后实现 WndProc，现在有声明就够了。接下来，需要一个应用程序入口点。Win32 应用程序和控制台应用程序的 main 函数原型稍有不同，如下代码所示：

```
int WINAPI WinMain(HINSTANCE hThis, HINSTANCE hPrev, LPSTR szCmdLine, int iCmdShow)
```

注意，在返回类型（int）后面有一个 WINAPI 宏，它表示一种调用约定（*calling convention*）。

WINAPI 或 stdcall 意味着栈的清理工作由被调函数来完成。WinMain 是函数名，该函数必须有 4 个参数，而且参数的顺序要与声明中的顺序相同。第 1 个参数 hThis 是应用程序当前实例的句柄。第 2 个参数 hPrev 是应用程序上一个实例的句柄。如果查阅 MSDN 文档（http://msdn.microsoft.com/en-us/library/windows/desktop/ms633559%28v=vs.85%29.aspx）可以看到，hPrev 参数一定是 NULL。我猜应该是为了兼容旧版本的 Windows 操作系统，所以没有写明当前版本的值。第 3 个参数是 szCmdLine 或应用程序的命令行，包括该程序的名称。最后一个参数控制如何显示窗口。

可以用 OR（|）运算符组合多个位值（欲了解详细内容，请参阅 MSDN）。

接下来，在 WinMain 的函数体中，用 UNREFERENCED_RARAMETER 宏告诉编译器不使用某些参数，方便编译器进行一些额外的优化。如下代码所示：

```
UNREFERENCED_PARAMETER( hPrev );
UNREFERENCED_PARAMETER( szCmdLine );
```

然后，实例化 WNDCLASSEX 窗口结构。该对象中储存了待生成窗口的细节，如栈大小、当前应用程序实例的句柄、窗口样式、窗口颜色、图标和鼠标指针。WNDCLASSEX 窗口结构的实例化代码如下所示：

```
WNDCLASSEX wndEx = { 0 };
```

下面的代码定义了在实例化窗口类后分配的额外字节数：

```
wndEx.cbClsExtra = 0;
```

下面的代码定义了窗口结构的大小（以字节为单位）：

```
wndEx.cbSize = sizeof( wndEx );
```

下面的代码定义了实例化窗口实例后分配的额外字节数：

```
wndEx.cbWndExtra = 0;
```

下面的代码定义了窗口类背景画刷的句柄：

```
wndEx.hbrBackground = (HBRUSH)(COLOR_WINDOW + 1);
```

下面的代码定义了窗口类光标的句柄：

```
wndEx.hCursor = LoadCursor( NULL, IDC_ARROW );
```

下面的代码定义了窗口类图标的句柄:

```
wndEx.hIcon = LoadIcon( NULL, IDI_APPLICATION );
wndEx.hIconSm = LoadIcon( NULL, IDI_APPLICATION );
```

下面的代码定义了包含窗口过程的实例句柄:

```
wndEx.hInstance = hThis;
```

下面的代码定义了指向窗口过程的指针:

```
wndEx.lpfnWndProc = WndProc;
```

下面的代码定义了指向以空字符结尾的字符串或原子的指针:

```
wndEx.lpszClassName = TEXT("GUIProject");
```

下面的代码定义了指向以空字符结尾的字符串的指针,该字符串指定了窗口类菜单的资源名:

```
wndEx.lpszMenuName = NULL;
```

下面的代码定义了窗口类的样式:

```
wndEx.style = CS_HREDRAW | CS_VREDRAW;
```

下面的代码注册一个窗口类,供 CreateWindow 或 CreateWindowEx 函数稍后使用:

```
if ( !RegisterClassEx( &wndEx ) )
{
    return -1;
}
```

CreateWindow API 创建一个重叠、弹出的窗口或子窗口。它指定该窗口类、窗口标题、窗口样式、窗口的初始位置和大小(可选的)。该函数还指定了窗口的父窗口或所有者(如果有的话),以及窗口的菜单。如下代码所示:

```
HWND hWnd = CreateWindow( wndEx.lpszClassName, TEXT("GUI Project"), WS_OVERLAPPEDWINDOW,
                          200, 200, 400, 300, HWND_DESKTOP,NULL, hThis, 0 );
if ( !hWnd )
{
    return -1;
}
```

如果指定窗口的更新域未被填满,UpdateWindow 函数就向窗口发送一条 WM_PAINT 消息,更新指定窗口的客户区。该函数绕过应用程序的消息队列,向指定窗口的窗口过程直接发送一条 WM_PAINT 消息。如下代码所示:

```
UpdateWindow( hWnd );
```

下面的代码设置指定窗口的显示状态:

```
ShowWindow( hWnd, iCmdShow );
```

我们还需要一个 MSG 结构的实例来表示窗口消息。

```
MSG msg = { 0 };
```

接下来，进入一个消息循环。Windows 中的应用程序是事件驱动的，它们不会显式调用函数（如，C 运行时库调用）来获得输入，而是等待系统把输入传递给它们。系统把所有的输入传递给应用程序的不同窗口。每个窗口都有一个叫做窗口过程的函数，当有输入需要传递给窗口时，系统调用会调用该函数。窗口过程处理输入，并把控制权返回系统。GetMessage API 从主调线程的消息队列中检索信息，如下代码所示：

```
while ( GetMessage( &msg, NULL, NULL, NULL ) )
{
    // 把虚拟键消息翻译成字符消息
    TranslateMessage(&msg );
    // 分发一条消息给窗口过程
    DispatchMessage(&msg );
}
```

当关闭应用程序或发送一些触发其退出的命令时，系统会释放应用程序消息队列。这意味着该应用程序不会再有消息，而且 while 循环也将结束。DestroyWindow API 销毁指定的窗口。该函数向指定窗口发送 WM_DESTROY 和 WM_NCDESTROY 消息，使窗口无效并移除其键盘焦点（*keyboard focus*）。此外，该函数还将销毁指定窗口的菜单，清空线程的消息队列，销毁与窗口过程相关的计时器，解除窗口对剪切板的所有权，如果该窗口在查看器链的顶端，还将打断剪切板的查看器链。

```
DestroyWindow( hWnd );
```

下面的函数注销窗口类，释放该类占用的内存：

```
UnregisterClass( wndEx.lpszClassName, hThis );
```

下面的 return 函数从应用程序消息队列中返回一个成功退出代码或最后一个消息代码，如下代码所示：

```
return (int) msg.wParam;
```

以上，我们逐行讲解了 WinMain 函数。接下来，要实现窗口过程或应用程序主事件处理器。作为第 1 个实例，先创建一个简单的 WndProc，它只有一个处理关闭窗口的功能。该窗口过程返回 64 位有符号长整型值，有 4 个参数：hWnd 结构（表示窗口标识符）、uMsg 无符号整数（表示窗口消息代码）、wParam 无符号 64 位长整型数（传递应用程序定义的数据）、lParam 有符号 64 位长整型数（也用于传递应用程序定义的数据）。

```
LRESULT CALLBACK WndProc( HWND hWnd, UINT uMsg, WPARAM wParam, LPARAM lParam )
{
```

消息代码负责处理消息，如默认消息（该例中是 WM_CLOSE），即正在关闭应用程序时系统发送的消息。然后，调用 PostQuitMessage API 释放系统资源，并安全关闭该应用程序。

```
    switch ( uMsg )
    {
        case WM_CLOSE:
        {
            PostQuitMessage( 0 );
            break;
        }
```

```
        default:
        {
```

最后，调用默认窗口过程（DefWindowProc）处理应用程序未处理的窗口消息。该函数确保每个消息都被处理，如下所示：

```
            return DefWindowProc(hWnd, uMsg, wParam, lParam);
        }
    }
    return 0;
}
```

虽然本节介绍的窗口应用程序示例非常简单，但是它完整地反映了事件驱动系统特性和事件处理机制。在后面的章节中，我们将频繁地使用事件处理，所以理解这些基本过程非常重要。

## 1.9 链表、队列和栈示例

下面的示例将演示线性链表（可包含任何泛型类型 T）的 OOP 用法。该示例背后的思想是把继承作为表示"B 是一种 A"这种关系的模型。

线性链表是一种线性排列元素的结构，第 1 个元素链接第 2 个元素，第 2 个元素链接第 3 个元素，以此类推。线性链表的基本操作是，在线性链表中插入元素（PUT）和获取元素（GET）。队列是一种线性链表，其中的元素按先进先出的次序排列，即 FIFO（*First In First Out*）。因此，从队列的顶部获取元素，从队列的底部插入新元素。栈也是一种线性链表，从栈中获取元素的顺序与放入元素的顺序相反，也就是说，栈中的元素按先进后出的次序排列，即 LIFO（*Last In First Out*）。

线性链表是按顺序排列的元素集合。每个链表都有用于设置元素值、从链表获取元素或简单查看元素的方法。链表可储存任何类型的对象。但是，为了满足链表类、队列和栈的特殊化，线性链表定义了插入元素和获取元素的精确位置。因此，作为泛化对象的链表是一个基类。

读者应该意识到，以这种方式设计的链表可实现为静态结构或动态结构。也就是说，这种链表可以实现为某种数组或结构，其中的各元素通过指针与相邻的元素链接。用指针来实现，可操作性强。

下面的示例中，将把线性链表实现为指针集合，这些指针指向用链表中的方法放置在链表中的原始对象。这样设计是为了实现链表元素的多态性，而不是把原始对象拷贝给链表。从语义上来看，这需要把更多的精力放在设计上。

### 准备就绪

确定安装并运行了 **Visual Studio**。

## 1.9 链表、队列和栈示例

### 操作步骤

执行以下步骤。

1. 创建一个新的空 C++ 控制台应用程序,名为 LinkedList。
2. 添加一个新的头文件 CList.h,并输入下面的代码:

```cpp
#ifndef _LIST_
#define _LIST_

#include <Windows.h>

template <class T>
class CNode
{
public:
    CNode(T* tElement) : tElement(tElement), next(0) { }
    T* Element() const { return tElement; }
    CNode*& Next(){ return next; }
private:
    T* tElement;
    CNode* next;
};

template <class T>
class CList
{
public:
    CList() : dwCount(0), head(0){ }
    CList(T* tElement) : dwCount(1), head(new CNode<T>(tElement)){ }
    virtual ~CList(){ }
    void Append(CNode<T>*& node, T* tElement);
    void Insert(T* tElement);
    bool Remove(T* tElement);
    DWORD Count() const { return dwCount; }
    CNode<T>*& Head() { return head; }
    T* GetFirst(){ return head != NULL ? head->Element() : NULL; }
    T* GetLast();
    T* GetNext(T* tElement);
    T* Find(DWORD(*Function)(T* tParameter), DWORD dwValue);
protected:
    CList(const CList& list);
    CList& operator = (const CList& list);
private:
    CNode<T>* head;
    DWORD dwCount;
};

template <class T>
void CList<T>::Append(CNode<T>*& node, T* tElement)
{
    if (node == NULL)
    {
        dwCount++;
        node = new CNode<T>(tElement);
        return;
    }
```

```cpp
        Append(node->Next(), tElement);
}

template <class T>
void CList<T>::Insert(T* tElement)
{
    dwCount++;
    if (head == NULL)
    {
        head = new CNode<T>(tElement);
        return;
    }
    CNode<T>* tmp = head;
    head = new CNode<T>(tElement);
    head->Next() = tmp;
}

template <class T>
bool CList<T>::Remove(T* tElement)
{
    if (head == NULL)
    {
        return NULL;
    }
    if (head->Element() == tElement)
    {
        CNode<T>* tmp = head;
        head = head->Next();

        delete tmp;
        dwCount--;

        return true;
    }

    CNode<T>* tmp = head;
    CNode<T>* lst = head->Next();

    while (lst != NULL)
    {
        if (lst->Element() == tElement)
        {
            tmp->Next() = lst->Next();

            delete lst;
            dwCount--;

            return true;
        }

        lst = lst->Next();
        tmp = tmp->Next();
    }

    return false;
}

template <class T>
T* CList<T>::GetLast()
```

```cpp
{
    if (head)
    {
        CNode<T>* tmp = head;
        while (tmp->Next())
        {
            tmp = tmp->Next();
        }
        return tmp->Element();
    }
    return NULL;
}

template <class T>
T* CList<T>::GetNext(T* tElement)
{
    if (head == NULL)
    {
        return NULL;
    }
    if (tElement == NULL)
    {
        return GetFirst();
    }
    if (head->Element() == tElement)
    {
        return head->Next() != NULL ? head->Next()->Element() : NULL;
    }

    CNode<T>* lst = head->Next();
    while (lst != NULL)
    {
        if (lst->Element() == tElement)
        {
            return lst->Next() != NULL ? lst->Next()->Element() : NULL;
        }

        lst = lst->Next();
    }

    return NULL;
}

template <class T>
T* CList<T>::Find(DWORD(*Function)(T* tParameter), DWORD dwValue)
{
    try
    {
        T* tElement = NULL;
        while (tElement = GetNext(tElement))
        {
            if (Function(tElement) == dwValue)
            {
                return tElement;
            }
        }
    }
    catch (...) {}
```

```
    return NULL;
}
#endif
```

3. 有了 CList 类的实现和定义，创建 CQueue 和 CStack 就很容易了。先创建 CQueue，右键单击【头文件】，创建一个新的头文件 CQueue.h，并输入下面的代码：

```
#ifndef __QUEUE__
#define __QUEUE__

#include "CList.h"

template<class T>
class CQueue : CList<T>
{
public:
    CQueue() : CList<T>(){ }
    CQueue(T* tElement) : CList<T>(tElement){ }
    virtual ~CQueue(){ }
    virtual void Enqueue(T* tElement)
    {
        Append(Head(), tElement);
    }
    virtual T* Dequeue()
    {
        T* tElement = GetFirst();
        Remove(tElement);
        return tElement;
    }
    virtual T* Peek()
    {
        return GetFirst();
    }
    CList<T>::Count;
protected:
    CQueue(const CQueue<T>& cQueue);
    CQueue<T>& operator = (const CQueue<T>& cQueue);
};

#endif
```

4. 类似地，再创建 CStack。右键单击【头文件】，创建一个新的头文件 CStack.h，并输入下面的以下代码：

```
#ifndef __STACK__
#define __STACK__

#include "CList.h"

template<class T>
class CStack : CList<T>
{
public:
    CStack() : CList<T>(){ }
    CStack(T* tElement) : CList<T>(tElement){ }
    virtual ~CStack(){ }
    virtual void Push(T* tElement)
```

```cpp
    {
        Insert(tElement);
    }
    virtual T* Pop()
    {
        T* tElement = GetFirst();
        Remove(tElement);
        return tElement;
    }
    virtual T* Peek()
    {
        return GetFirst();
    }
    CList<T>::Count;
protected:
    CStack(const CStack<T>& cStack);
    CStack<T>& operator = (const CStack<T>& cStack);
};

#endif
```

5. 最后, 实现 LinkedList.cpp 中的代码, 我们用来充当 main 例程:

```cpp
#include <iostream>

using namespace std;

#include "CQueue.h"
#include "CStack.h"

int main()
{
    CQueue<int>* cQueue = new CQueue<int>();
    CStack<double>* cStack = new CStack<double>();

    for (int i = 0; i < 10; i++)
    {
        cQueue->Enqueue(new int(i));
        cStack->Push(new double(i / 10.0));
    }

    cout << "Queue - integer collection:" << endl;
    for (; cQueue->Count();)
    {
        cout << *cQueue->Dequeue() << " ";
    }

    cout << endl << endl << "Stack - double collection:" << endl;
    for (; cStack->Count();)
    {
        cout << *cStack->Pop() << " ";
    }

    delete cQueue;
    delete cStack;

    cout << endl << endl;
    return system("pause");
}
```

## 示例分析

首先，解释一下 CList 类。为了更方便地处理，该链表由 CNode 类型的元素组成。CNode 类有两个特征：tElement 指针（指向用户自定义的元素）和 next 指针（指向链表下一个项）；实现了两个方法：Element 和 Next。Element 方法返回当前元素地址的指针，Next 方法返回下一个项地址的引用。

从文字方面看，构造函数称为 ctor，析构函数称为 dtor。CList 的默认构造函数是公有函数，创建一个空的链表。第 2 个构造函数创建一个包含一个开始元素的链表。具有动态结构的链表必须有析构函数。Append 方法在链表的末尾插入一个元素，也是链表的最后一个元素。Count 方法返回链表当前的元素个数。Head 方法返回链表开始节点的引用。

GetFirst 方法返回链表的第 1 个元素，如果链表为空，则返回 NULL。GetLast 方法返回链表的最后一个元素，如果链表为空，则返回 NULL。GetNext 方法返回链表的下一个项，即相对于地址由 T* tElement 参数提供的项的下一个项。如果未找到该项，GetNext 方法返回 NULL。

Find 方法显然是一个高级特性，针对未定义类型 T 和未定义的 Function 方法（带 tParameter 参数）设计。假设要使用一个包含学生对象的链表，例如迭代数据（如，使用 GetNext 方法）或查找某个学生。如果有可能为将来定义的类型实现一个返回 unsigned long 类型（DWORD）的方法，而且该方法要把未知类型数据与 dwValue 参数做比较，应该怎么做？例如，假设要根据学生的 ID 找出这名学生，可以使用下面的代码：

```cpp
#include <windows.h>
#include "CList.h"

class CStudent
{
public:
    CStudent(DWORD dwStudentId) : dwStudentId(dwStudentId){ }
    static DWORD GetStudentId(CStudent* student)
    {
        DWORD dwValue = student->GetId();
        return dwValue;
    }
    DWORD GetId() const
    {
        return dwStudentId;
    }
private:
    DWORD dwStudentId;
};

int main()
{
    CList<CStudent>* list = new CList<CStudent>();
    list->Insert(new CStudent(1));
    list->Insert(new CStudent(2));
    list->Insert(new CStudent(3));
    CStudent* s = list->Find(&CStudent::GetStudentId, 2);
```

```
        if (s != NULL)
        {
            // 找到 s
        }

        return 0;
    }
```

如果链表用于处理基本类型（如，int），可使用下面的代码：

```
#include <windows.h>
#include "CList.h"

DWORD Predicate(int* iElement)
{
    return (DWORD)(*iElement);
}

int main()
{
    CList<int>* list = new CList<int>();
    list->Insert(new int(1));
    list->Insert(new int(2));
    list->Insert(new int(3));

    int* iElement = list->Find(Predicate, 2);
    if (iElement != NULL)
    {
        // 找到 iElement
    }
    return 0;
}
```

回到我们的示例。为何要把拷贝构造函数和 operator= 都声明为 protected？要知道，我们在这里实现的链表中储存着指针，而这些指针指向那些在链表外的对象。如果用户能随意（或无意）地通过拷贝构造函数或=运算符来拷贝链表，会非常危险。因此，要把拷贝构造函数和=运算符设置为 protected，不让用户使用。为此，必须先把两个函数声明为 protected，然后在需要时再实现它们；否则，编译器在默认情况下会把这两个函数设置为 public，然后逐个拷贝指针，这样做是错误的。把它们声明为 private 也不够。在这种情况下，CList 基类的派生类依旧会遇到同样的问题。派生类仍需要要把拷贝构造函数和=运算符声明为 protected，否则编译器还是会把这些方法默认生成 public。如果基类包含拷贝构造函数和=运算符，派生类就会默认调用它们，除非派生类能显式调用自己的版本。但是，我们的初衷是让派生类用上 CList 基类中的拷贝构造函数和=运算符，所以将其声明为 protected。

CList 类的 private 部分包含了把链表实现为线性链表所需的对象。这意味着链表中的每个元素都指向下一个元素，头节点指向第 1 个元素。

CQueue 和 CStack 类分别实现为队列和栈。不难看出，设计好 CList 基类以后（尤其是设计了 Enqueue、Dequeue、Push、Pop 和 Peek 方法），实现这两个类有多么简单。只要 CList 类设计得当，设计 CQueue、CStack，甚至其他类都非常简单。

# 第 2 章 进程和线程的概念

**本章介绍以下内容：**
- 进程和线程
- 解释进程模型
- 进程的实现
- 进程间通信（IPC）
- 解决典型的 IPC 问题
- 线程模型的实现
- 线程的用法
- 在用户空间实现线程
- 在内核实现线程

## 2.1 简介

现在的计算机能同时处理多件事，许多 Windows 用户还没有完全意识到这一点。我们举例说明一下。当启动 PC 系统时，许多进程都在后台启动（例如，管理电子邮件的进程、负责更新病毒库的进程等）。通常，用户在执行其他任务时（如，上网），还会打印文件或播放 CD。这些活动都需要管理。支持多进程的多任务系统处理这些情况得心应手。在这种多任务系统中，CPU 以极快的速度在各进程间切换，每个进程仅运行几毫秒。从严格意义上来说，CPU 在任何时刻只运行一个进程，只不过它快速切换进程营造了并行处理的假象。

近些年来，操作系统演变为一个顺序的概念模型（顺序进程）。包括操作系统在内，所有的可运行软件都在计算机中表现为一系列顺序进程。进程是执行程序的实例。每个进程都有自己的虚拟地址空间和控制线程。线程是操作系统**调度器**（*scheduler*）分配处理器时间的基础单元。我们可以把系统看作是运行在准并行环境中的进程集合。在进程（程序）间快速地反复切换叫做多任务处理。

## 2.2 进程和线程

图 2.1 演示了执行 4 个程序调度的一个单核 CPU 多任务处理系统。图 2.2 演示了执行 4 个进程的一个多核 CPU 多任务处理系统，每个进程单独运行，各有一个控制流。

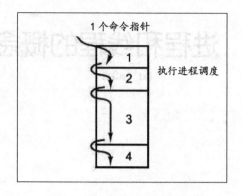
图 2.1 单核 CPU 多任务处理系统

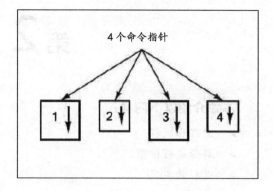
图 2.2 多核 CPU 多任务处理系统

如图 2.3 所示，随着时间的推移，虽然进程有不同程度的进展，但是在每一时刻，单核 CPU 只运行一个进程。

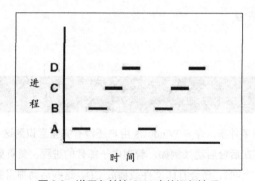
图 2.3 进程在单核 CPU 中的运行情况

前面提到过，在传统的操作系统中，每个进程都有一个地址空间和一个控制线程。在许多情况下，一个进程的地址空间中要执行多个线程，在准并行上下文中，这些线程就像是不同的进程一样。有多个线程的主要原因是，许多应用程序都要求能立即执行多项操作。当然，某些操作可以等待（阻塞）一段时间。把运行在准并行上下文中的应用程序分解成多个单独的线程，程序设计模型就变得更简单了。通过添加线程，操作系统提供了一个新特性：并行实体能共享一个地址空间和它们的所有数据。这是执行并发的必要条件。

## 2.3 解释进程模型

传统的操作系统必须提供创建进程和终止进程的方法。下面列出了 4 个引发创建进程的主要事件：

- 系统初始化；
- 正在运行的进程执行创建进程的系统调用；

## 2.3 解释进程模型

- 用户要求创建新进程；
- 启动批处理作业。

操作系统启动后，会创建多个进程。一些是前台进程，与用户（人）交互，并根据用户的要求执行操作。一些是后台进程，执行特定的功能，与用户行为不相关。例如，可以把接收电子邮件设计成后台进程，因为大部分时间都用不到这一功能，只需在有电子邮件到达时处理即可。后台进程通常处理诸如电子邮件、打印等活动。

在 Windows 中，CreateProcess（一个 Win32 函数调用）负责创建进程和加载进程上下文。欲详细了解 CreateProcess，请参阅 MSDN（http://msdn.microsoft.com/en-us/library/windows/desktop/ ms682425%28v=vs.85%29.aspx）。我们将在下面的示例中演示基本的进程创建和同步。

### 准备就绪

确定安装并运行了 Visual Studio。

### 操作步骤

现在，我们按下面的步骤创建一个程序，稍后再详细解释。

1. 创建一个新的默认 C++ 控制台应用程序，名为 ProcessDemo。

2. 打开 ProcessDemo.cpp。

3. 添加下面的代码：

```cpp
#include "stdafx.h"
#include <Windows.h>
#include <iostream>

using namespace std;

int _tmain(int argc, _TCHAR* argv[])
{
    STARTUPINFO startupInfo = { 0 };
    PROCESS_INFORMATION processInformation = { 0 };

    BOOL bSuccess = CreateProcess(
        TEXT("C:\\Windows\\notepad.exe"), NULL, NULL,
        NULL, FALSE, NULL, NULL, NULL, &startupInfo,
        &processInformation);

    if (bSuccess)
    {
        cout << "Process started." << endl
            << "Process ID:\t"
            << processInformation.dwProcessId << endl;
    }
    else
```

```
            {
                cout << "Cannot start process!" << endl
                    << "Error code:\t" << GetLastError() << endl;
            }
            return system("pause");
        }
```

该程序的输出如图 2.4 所示。

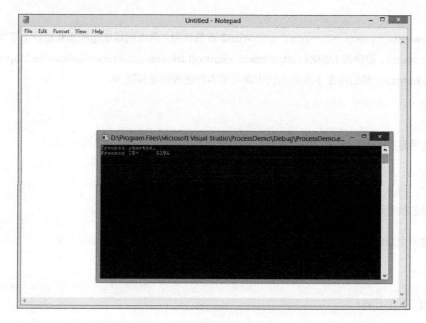

图 2.4

如图 2.4 所示,开始了一个新的进程(记事本)。

## 示例分析

CreateProcess 函数用于创建一个新进程及其主线程。新进程在主调进程的安全上下文中运行。

操作系统为进程分配进程标识符。进程标识符用于标识进程,在进程终止之前有效。或者,对于一些 API(如 OpenProcess 函数),进程标识符用于获得进程的句柄。最初线程的线程标识符由进程分配,该标识符可用于在 OpenThread 中打开一个线程的句柄。线程标识符在线程终止之前有效,可作为系统中线程的唯一标识。这些标识符都返回 PROCESS_INFORMATION 结构中。

主调线程可以使用 WaitForInputIdle 函数,在新进程完成其初始化且正在等待用户输入时等待。这在父进程和子进程的同步中很有用,因为 CreateProcess 不会等到新进程完成初始化才返回。例如,创建的进程会在查找与新进程相关的窗口之前使用 WaitForInputIdle 函数。

终止进程较好的做法是调用 `ExitProcess` 函数,因为它会给所属进程的所有 DLL 都发送一条即将终止的通知。而关闭进程的其他方法就不会这样做(如,`TerminateProcess` API)。注意,只要进程中有一个线程调用 `ExitProcess` 函数,该进程的其他线程都会立即终止,根本没机会执行其他代码(包括相关 DLL 的线程终止代码)。

## 更多讨论

虽然每个进程都是一个独立的实体,有各自的指令指针和内部状态,但是进程之间也要经常交互。一个进程生成的输出数据可能是另一个进程所需的输入数据。根据两个进程的相对运行速度,可能会发生这种情况:读操作已准备运行,但是却没有输入。在能读到输入数据之前,该进程必定被阻塞。从逻辑上看,如果进程被阻塞就不能继续运行,因为该进程正在等待尚未获得的输入。如果操作系统在这时决定把 CPU 暂时分配给另一个进程,正在等待的进程就有可能停止。这是两种完全不同的情况。第一种情况是问题本身造成的(即,在用户键入数据之前无法解析用户的命令行);而第二种情况是系统的技术原因造成的(即,进程用完了分配给它的时间,又没有足够的 CUP 能单独运行该线程)。

图 2.5 中的状态图演示了一个进程可能处于的 3 种状态。

- ▶ **运行**:此时,该进程正在使用 CPU。
- ▶ **就绪**:该进程可运行,但是它暂时停止让其他进程运行。
- ▶ **阻塞**:在某些外部事件发生前,该进程不能运行。

**运行**和**就绪**状态有些类似。处于这两种状态的进程都可以运行,只是在就绪状态中,进程暂时没有 CPU 可用。**阻塞**状态与前两种状态不同,在阻塞状态中,即使 CPU 空闲,进程也不能运行。

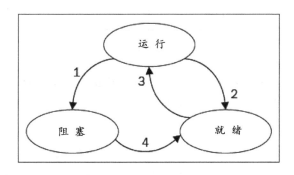

图 2.5 进程的三种状态

假设有进程 A。当调度器选择另一个进程在 CPU 上执行时,进程 A 发生转换过程 2。当调度器选择执行进程 A 时,发生转换过程 3。为了等待输入,转换过程 1 把进程 A 设置为阻塞状态。当进程 A 获得所需的输入时,发生转换过程 4。

## 2.4 进程的实现

在现代的多任务系统中，**进程控制块**（*Process Control Block*，PCB）储存了高效管理进程所需的许多不同数据项。PCB 是操作系统为了管理进程，在内核中设置的一种的数据结构。操作系统中的进程用 PCB 来表示。虽然这种数据结构的细节因系统而异，但是常见的部分大致可分为 3 大类：

- 进程标识数据；
- 进程状态数据；
- 进程控制数据。

图 2.6

PCB 是管理进程的中心。绝大多数操作系统程序（包括那些与调度、内存、I/O 资源访问和性能监控相关的程序）都要访问和修改它。通常，要根据 PCB 为进程构建数据。例如，某 PCB 内指向其他 PCB 的指针以不同的调度状态（就绪、阻塞等）创建进程队列。

操作系统必须代表进程来管理资源。它必须不断地关注每个进程的状态、系统资源和内部值。下面的程序示例演示了如何获得一个进程基本信息结构地址，其中的一个特征就是 PCB 的地址。另一个特征是唯一的进程 ID。为简化示例，我们只输出从对象中读取的进程 ID。

### 准备就绪

确定安装并运行了 Visual Studio。

### 操作步骤

我们再来创建一个操控进程的程序。这次,我们从进程基本信息结构中获取进程 ID。请执行以下步骤。

1. 创建一个新的默认 C++ 控制台应用程序,名为 NtProcessDemo。

2. 打开 NtProcessDemo.cpp。

3. 添加下面的代码:

```
#include "stdafx.h"
#include <Windows.h>
#include <Winternl.h>
#include <iostream>

using namespace std;

typedef NTSTATUS(NTAPI* QEURYINFORMATIONPROCESS)(
    IN HANDLE ProcessHandle,
    IN PROCESSINFOCLASS ProcessInformationClass,
    OUT PVOID ProcessInformation,
    IN ULONG ProcessInformationLength,
    OUT PULONG ReturnLength OPTIONAL
    );

int _tmain(int argc, _TCHAR* argv[])
{
    STARTUPINFO startupInfo = { 0 };
    PROCESS_INFORMATION processInformation = { 0 };

    BOOL bSuccess = CreateProcess(
        TEXT("C:\\Windows\\notepad.exe"), NULL, NULL,
        NULL, FALSE, NULL, NULL, NULL, &startupInfo,
        &processInformation);

    if (bSuccess)
    {
        cout << "Process started." << endl << "Process ID:\t"
            << processInformation.dwProcessId << endl;

        PROCESS_BASIC_INFORMATION pbi;
        ULONG uLength = 0;

        HMODULE hDll = LoadLibrary(
            TEXT("C:\\Windows\\System32\\ntdll.dll"));

        if (hDll)
        {
            QEURYINFORMATIONPROCESS QueryInformationProcess =
                (QEURYINFORMATIONPROCESS)GetProcAddress(
                    hDll, "NtQueryInformationProcess");

            if (QueryInformationProcess)
            {
                NTSTATUS ntStatus = QueryInformationProcess(
                    processInformation.hProcess,
                    PROCESSINFOCLASS::ProcessBasicInformation,
```

```
                        &pbi, sizeof(pbi), &uLength);
                    if (NT_SUCCESS(ntStatus))
                    {
                        cout << "Process ID (from PCB):\t"
                            << pbi.UniqueProcessId << endl;
                    }
                    else
                    {
                        cout << "Cannot open PCB!" << endl
                            << "Error code:\t" << GetLastError()
                            << endl;
                    }
                }
                else
                {
                    cout << "Cannot get "
                        << "NtQueryInformationProcess function!"
                        << endl << "Error code:\t"
                        << GetLastError() << endl;
                }
                FreeLibrary(hDll);
            }
            else
            {
                cout << "Cannot load ntdll.dll!" << endl
                    << "Error code:\t" << GetLastError() << endl;
            }
        }
        else
        {
            cout << "Cannot start process!" << endl
                << "Error code:\t" << GetLastError() << endl;
        }
        return 0;
    }
```

## 示例分析

该例中，我们使用了一些其他头文件：Winternl.h 和 Windows.h。Winternl.h 头文件包含了大部分 Windows 内部函数的原型和数据表示，例如 PROCESS_BASIC_INFORMATION 结构的定义：

```
    typedef struct _PROCESS_BASIC_INFORMATION {
        PVOID Reserved1;
        PPEB PebBaseAddress;
        PVOID Reserved2[2];
        ULONG_PTR UniqueProcessId;
        PVOID Reserved3;
    } PROCESS_BASIC_INFORMATION;
```

操作系统在调用内核态和用户态之间的子例程时会用到该结构。

结合 PROCESSINFOCLASS::ProcessBasicInformation 枚举，我们通过 UniqueProcessId 特征获取进程标识符，如上面的代码所示。

首先，定义 QEURYINFORMATIONPROCESS，这是从 ntdll.dll 中加载的 NtQueryInformationProcess

函数的别名。当通过 `GetProcAddress` Win32 API 获得该函数的地址时，就可以询问 `PROCESS_BASIC_INFORMATION` 对象了。注意 `PROCESS_BASIC_INFORMATION` 结构的 `PebBaseAddress` 字段是一个指针，指向新创建进程的 PCB。如果还想进一步研究 PCB，检查新创建的进程，则必须在运行时使用 `ReadProcessMemory` 例程。因为 `PebBaseAddress` 指向属于新创建进程的内存。

## 2.5 进程间通信（IPC）

进程之间的通信非常重要。虽然操作系统提供了进程间通信的机制，但是在介绍这些机制之前，我们先来考虑一些与之相关的问题。如果航空预定系统中有两个进程在同时销售本次航班的最后一张机票，怎么办？这里要解决两个问题。第 1 个问题是，一个座位不可能卖两次。第 2 个问题是一个依赖性问题：如果进程 A 生成的某些数据是进程 B 需要读取的（如，打印这些数据），那么进程 B 在进程 A 准备好这些数据之前必须一直等待。进程和线程的不同在于，线程共享同一个地址空间，而进程拥有单独的地址空间。因此，用线程解决第 1 个问题比较容易。至于第 2 个问题，线程也同样能解决。所以，理解同步机制非常重要。

在讨论 IPC 之前，我们先来考虑一个简单的例子：CD 刻录机。当一个进程要刻录一些内容时，会在特定的刻录缓冲区中设置文件句柄（我们立刻要刻录更多的文件）。另一个负责刻录的进程，检查待刻录的文件是否存在，如果存在，该进程将刻录文件，然后从缓冲区中移除该文件的句柄。假设刻录缓冲区有足够多的索引，分别编号为 I0、I1、I2 等，每个索引都能储存若干文件句柄。再假设有两个共享变量：`p_next` 和 `p_free`，前者指向下一个待刻录的缓冲区索引，后者指向缓冲区中的下一个空闲索引。所有进程都要使用这两个变量。在某一时刻，索引 I0 和 I2 为空（即文件已经刻录完毕），I3 和 I5 已经加入缓冲。同时，**进程 5** 和**进程 6** 决定把文件句柄加入队列准备刻录文件。这一状况如图 2.7 所示。

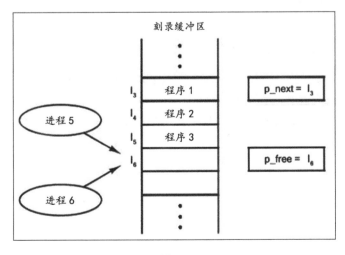

图 2.7

首先，**进程** 5 读取 `p_free`，把它的值 I6 储存在自己的局部变量 `f_slot` 中。接着，发生了一个时钟中断，CPU 认为**进程** 5 运行得太久了，决定转而执行**进程** 6。然后，**进程** 6 也读取 `p_free`，同样也把 I6 储存在自己的局部变量 `f_slot` 中。此时，两个进程都认为下一个可用的索引是 I6。**进程** 6 现在继续运行，它把待拷贝文件的句柄储存在索引 I6 中，并更新 `p_free` 为 I7。然后，系统让**进程** 6 睡眠。现在，**进程** 5 从原来暂停的地方再次开始运行。它查看自己的 `f_slot`，发现可用的索引是 I6，于是把自己待拷贝文件的句柄写到索引 I6 上，擦除了**进程** 6 刚写入的文件句柄。然后，**进程** 5 计算 `f_slot+1` 得 I7，就把 `p_free` 设置为 I7。现在，刻录缓冲区内部保持一致，所以刻录进程并未出现任何错误。但是，**进程** 6 再也接收不到任何输出。

**进程** 6 将被无限闲置，等待着再也不会有的输出。像这样两个或更多实体读取或写入某共享数据的情况，最终的结果取决于进程的执行顺序（即何时执行哪一个进程），这叫做**竞态条件**（*race condition*）。

如何避免竞态条件？大部分解决方案都涉及共享内存、共享文件以及避免不同的进程同时读写共享数据。换句话说，我们需要**互斥**（*mutual exclusion*）或一种能提供独占访问共享对象的机制（无论它是共享变量、共享文件还是其他对象）。当**进程** 6 开始使用**进程** 5 刚用完的一个共享对象时，就会发生糟糕的事情。

程序中能被访问共享内存的部分叫做**临界区**（*critical section*）。为了避免竞态条件，必须确保一次只能有一个进程进入临界区。这种方法虽然可以避免竞态条件，但是在执行并行进程时会影响效率，毕竟并行的目的是正确且高效地合作。要使用共享数据，必须处理好下面 4 个条件：

- 不允许同时有两个进程在临界区内；
- 不得对 CPU 的速度或数量进行假设；
- 在临界区外运行的进程不得阻碍其他进程；
- 不得有任何进程处于永远等待进入临界区。

以上所述如图 2.8 所示。过程 A 在 $T_1$ 时进入临界区。稍后，进程 B 在 $T_2$ 尝试进入其临界区，但是失败。因为另一个进程已经在临界区中，同一时间内只允许一个进程在临界区内。在 $T_3$ 之前，进程 B 必须被临时挂起。在进程 A 离开临界区时，进程 B 便可立即进入。最终，进程 B 离开临界区（$T_4$ 时），又回到没有进程进入临界区的状态。

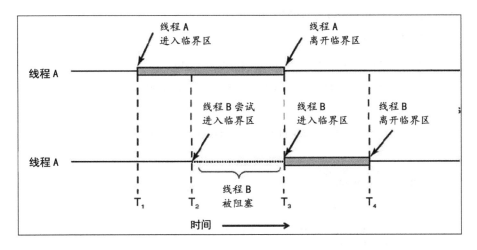

图 2.8

下面是一个进程间通信的程序示例。我们创建的这个程序一开始就有两个进程，它们要在一个普通窗口中完成绘制矩形的任务。从某种程度上看，这两个进程需要相互通信，即当一个进程正在画矩形时，另一个进程要等待。

### 准备就绪

确定安装并运行了 Visual Studio。

### 操作步骤

1. 创建一个新的默认 C++ 控制台应用程序，命名为 `IPCDemo`。

2. 右键单击【解决方案资源管理器】，并选择【添加】-【新建项目】。选择 C++【Win32 控制台应用程序】，添加一个新的默认 C++ 控制台应用程序，命名为 `IPCWorker`。

3. 在 `IPCWorker.cpp` 文件中输入下面的代码：

   ```
   #include "stdafx.h"
   #include <Windows.h>

   #define COMMUNICATION_OBJECT_NAME TEXT("__FILE_MAPPING__")
   #define SYNCHRONIZING_MUTEX_NAME TEXT( "__TEST_MUTEX__" )

   typedef struct _tagCOMMUNICATIONOBJECT
   {
       HWND hWndClient;
       BOOL bExitLoop;
       LONG lSleepTimeout;
   } COMMUNICATIONOBJECT, *PCOMMUNICATIONOBJECT;
   ```

```cpp
int _tmain(int argc, _TCHAR* argv[])
{
    HBRUSH hBrush = NULL;

    if (_tcscmp(TEXT("blue"), argv[0]) == 0)
    {
        hBrush = CreateSolidBrush(RGB(0, 0, 255));
    }
    else
    {
        hBrush = CreateSolidBrush(RGB(255, 0, 0));
    }

    HWND hWnd = NULL;
    HDC hDC = NULL;
    RECT rectClient = { 0 };
    LONG lWaitTimeout = 0;
    HANDLE hMapping = NULL;
    PCOMMUNICATIONOBJECT pCommObject = NULL;
    BOOL bContinueLoop = TRUE;

    HANDLE hMutex = OpenMutex(MUTEX_ALL_ACCESS, FALSE, SYNCHRONIZING_MUTEX_NAME);
    hMapping = OpenFileMapping(FILE_MAP_READ, FALSE, COMMUNICATION_OBJECT_NAME);

    if (hMapping)
    {
        while (bContinueLoop)
        {
            WaitForSingleObject(hMutex, INFINITE);
            pCommObject = (PCOMMUNICATIONOBJECT)MapViewOfFile(hMapping,
                FILE_MAP_READ, 0, 0, sizeof(COMMUNICATIONOBJECT));

            if (pCommObject)
            {
                bContinueLoop = !pCommObject->bExitLoop;
                hWnd = pCommObject->hWndClient;
                lWaitTimeout = pCommObject->lSleepTimeout;
                UnmapViewOfFile(pCommObject);
                hDC = GetDC(hWnd);
                if (GetClientRect(hWnd, &rectClient))
                {
                    FillRect(hDC, &rectClient, hBrush);
                }

                ReleaseDC(hWnd, hDC);
                Sleep(lWaitTimeout);
            }
            ReleaseMutex(hMutex);
        }
    }

    CloseHandle(hMapping);
    CloseHandle(hMutex);
    DeleteObject(hBrush);

    return 0;
}
```

4. 打开 IPCDemo.cpp,并输入下面的代码:

```cpp
#include "stdafx.h"
#include <Windows.h>
#include <iostream>

using namespace std;

#define COMMUNICATION_OBJECT_NAME TEXT("__FILE_MAPPING__")
#define SYNCHRONIZING_MUTEX_NAME TEXT( "__TEST_MUTEX__" )
#define WINDOW_CLASS_NAME TEXT( "__TMPWNDCLASS__" )
#define BUTTON_CLOSE 100

typedef struct _tagCOMMUNICATIONOBJECT
{
    HWND hWndClient;
    BOOL bExitLoop;
    LONG lSleepTimeout;
} COMMUNICATIONOBJECT, *PCOMMUNICATIONOBJECT;

LRESULT CALLBACK WndProc(HWND hDlg, UINT uMsg, WPARAM wParam, LPARAM lParam);
HWND InitializeWnd();
PCOMMUNICATIONOBJECT pCommObject = NULL;
HANDLE hMapping = NULL;

int _tmain(int argc, _TCHAR* argv[])
{
    cout << "Interprocess communication demo." << endl;
    HWND hWnd = InitializeWnd();
    if (!hWnd)
    {
        cout << "Cannot create window!" << endl << "Error:\t" <<
            GetLastError() << endl;
        return 1;
    }
    HANDLE hMutex = CreateMutex(NULL, FALSE, SYNCHRONIZING_MUTEX_NAME);
    if (!hMutex)
    {
        cout << "Cannot create mutex!" << endl << "Error:\t" <<
            GetLastError() << endl;
        return 1;
    }
    hMapping = CreateFileMapping((HANDLE)-1, NULL, PAGE_READWRITE, 0,
            sizeof(COMMUNICATIONOBJECT), COMMUNICATION_OBJECT_NAME);
    if (!hMapping)
    {
        cout << "Cannot create mapping object!" << endl << "Error:\t"
            << GetLastError() << endl;
        return 1;
    }
    pCommObject = (PCOMMUNICATIONOBJECT)MapViewOfFile(hMapping,
        FILE_MAP_WRITE, 0, 0, 0);
    if (pCommObject)
    {
        pCommObject->bExitLoop = FALSE;
        pCommObject->hWndClient = hWnd;
        pCommObject->lSleepTimeout = 250;
        UnmapViewOfFile(pCommObject);
    }

    STARTUPINFO startupInfoRed = { 0 };
```

```cpp
            PROCESS_INFORMATION processInformationRed = { 0 };
            STARTUPINFO startupInfoBlue = { 0 };
            PROCESS_INFORMATION processInformationBlue = { 0 };

            BOOL bSuccess = CreateProcess(TEXT("..\\Debug\\IPCWorker.exe"),
                TEXT("red"), NULL, NULL, FALSE, 0, NULL, NULL, &startupInfoRed,
                &processInformationRed);
            if (!bSuccess)
            {
                cout << "Cannot create process red!" << endl << "Error:\t" <<
                    GetLastError() << endl;
                return 1;
            }
            bSuccess = CreateProcess(TEXT("..\\Debug\\IPCWorker.exe"),
                TEXT("blue"), NULL, NULL, FALSE, 0, NULL, NULL, &startupInfoBlue,
                &processInformationBlue);
            if (!bSuccess)
            {
                cout << "Cannot create process blue!" << endl << "Error:\t" <<
                    GetLastError() << endl;
                return 1;
            }
            MSG msg = { 0 };
            while (GetMessage(&msg, NULL, 0, 0))
            {
                TranslateMessage(&msg);
                DispatchMessage(&msg);
            }
            UnregisterClass(WINDOW_CLASS_NAME, GetModuleHandle(NULL));
            CloseHandle(hMapping);
            CloseHandle(hMutex);
            cout << "End program." << endl;
            return 0;
        }

        LRESULT CALLBACK WndProc(HWND hWnd, UINT uMsg, WPARAM wParam, LPARAM lParam)
        {
            switch (uMsg)
            {
                case WM_COMMAND:
                {
                    switch (LOWORD(wParam))
                    {
                        case BUTTON_CLOSE:
                        {
                            PostMessage(hWnd, WM_CLOSE, 0, 0);
                            break;
                        }
                    }
                    break;
                }
                case WM_DESTROY:
                {
                    pCommObject = (PCOMMUNICATIONOBJECT)MapViewOfFile(hMapping,
                        FILE_MAP_WRITE, 0, 0, 0);
                    if (pCommObject)
                    {
                        pCommObject->bExitLoop = TRUE;
                        UnmapViewOfFile(pCommObject);
```

```
            }
            PostQuitMessage(0);
            break;
        }
        default:
        {
            return DefWindowProc(hWnd, uMsg, wParam, lParam);
        }
    }
    return 0;
}

HWND InitializeWnd()
{
    WNDCLASSEX wndEx;
    wndEx.cbSize = sizeof(WNDCLASSEX);
    wndEx.style = CS_HREDRAW | CS_VREDRAW;
    wndEx.lpfnWndProc = WndProc;
    wndEx.cbClsExtra = 0;
    wndEx.cbWndExtra = 0;
    wndEx.hInstance = GetModuleHandle(NULL);
    wndEx.hbrBackground = (HBRUSH)(COLOR_WINDOW + 1);
    wndEx.lpszMenuName = NULL;
    wndEx.lpszClassName = WINDOW_CLASS_NAME;
    wndEx.hCursor = LoadCursor(NULL, IDC_ARROW);
    wndEx.hIcon = LoadIcon(wndEx.hInstance, MAKEINTRESOURCE(IDI_APPLICATION));
    wndEx.hIconSm = LoadIcon(wndEx.hInstance, MAKEINTRESOURCE(IDI_APPLICATION));
    if (!RegisterClassEx(&wndEx))
    {
        return NULL;
    }
    HWND hWnd = CreateWindow(wndEx.lpszClassName,
        TEXT("Interprocess communication Demo"),
        WS_OVERLAPPEDWINDOW, 200, 200, 400, 300, NULL, NULL,
        wndEx.hInstance, NULL);
    if (!hWnd)
    {
        return NULL;
    }
    HWND hButton = CreateWindow(TEXT("BUTTON"), TEXT("Close"),
        WS_CHILD | WS_VISIBLE | BS_PUSHBUTTON | WS_TABSTOP,
        275, 225, 100, 25, hWnd, (HMENU)BUTTON_CLOSE, wndEx.hInstance,
        NULL);
    HWND hStatic = CreateWindow(TEXT("STATIC"), TEXT(""), WS_CHILD |
        WS_VISIBLE, 10, 10, 365, 205, hWnd, NULL, wndEx.hInstance, NULL);
    ShowWindow(hWnd, SW_SHOW);
    UpdateWindow(hWnd);

    return hStatic;
}
```

## 示例分析

这次演示的示例有点难。我们需要两个单独的线程，所以在同一个解决方案中创建了两个项目。

为了简化这个示例，我们在主应用程序 IPCDemo 中创建了两个进程，IPCDemo 将在应用程序窗口中绘制一个区域。如果没有正确的通信和进程同步，就会发生多路访问共享资源的情况。考虑到操作系统会在进程间快速切换，而且大部分 PC 都有多核 CPU，这很可能会导致两个进程同时画一个区域，即多个进程同时访问未保护的区域。先来看 IPCWorker，这个名称的意思是，需要进程为我们处理一些工作。

我们使用了一个映射对象（即，内存中为进程分配读取或写入的区域）。IPCWorker 或简称 Worker，要请求获得一个已命名的互斥量。如果获得互斥量，该进程就能处理并获取一个指向内存区域（文件映射）的指针，信息将储存在这个区域。必须获得互斥量，才能进行独占访问。进程在 WaitForSingleObject 返回后获得互斥量。请看下面的语句：

```
HANDLE hMutex = OpenMutex( MUTEX_ALL_ACCESS, FALSE, SYNCHRONIZING_MUTEX_NAME );
```

我们要为互斥量（hMutex）分配一个句柄，调用 OpenMutex Win32 API 获得该已命名互斥量的句柄（如果有互斥量的话）。请看下面的语句：

```
WaitForSingleObject( hMutex, INFINITE );
```

执行完这条语句后，当 WaitForSingleObject API 返回时继续执行。

```
pCommObject = ( PCOMMUNICATIONOBJECT )
    MapViewOfFile( hMapping, FILE_MAP_READ, 0, 0, sizeof( COMMUNICATIONOBJECT ) );
```

调用 MapViewOfFile Win32 API 获得指向文件映射对象的句柄（指针）。现在，进程可以从共享内存对象中读取并获得所需的信息了。该进程要读取 bExitLoop 变量才能获悉是否继续执行。然后，该进程要读取待绘制区域窗口的句柄（hWnd）。最后，还需要 lSleepTimeout 变量记录进程睡眠多久。我们故意添加了 sleep 时间，因为进程间切换太快根本注意不到。

```
ReleaseMutex( hMutex );
```

调用 ReleaseMutex Win32 API 释放互斥量的所有权，让其他进程可以获得互斥量，继续执行其他任务。分析完 IPCWorker，我们来看 IPCDemo 项目。该项目定义了 _tagCOMMUNICATIONOBJECT 结构，用于整个文件映射过程中对象之间的通信。

文件映射（*file mapping*）是把文件的内容与一个进程的一部分虚拟地址空间相关联。操作系统创建一个文件映射对象（也叫做区域对象[*section object*]）来维护这种关联。文件视图（*file view*）是进程用于访问文件内容的虚拟地址空间部分。有了文件映射，进程不仅能使用随机 I/O 和顺序 I/O，而且无需把整个文件映射到内存中就能高效地使用大型数据文件（如，数据库）。多个进程还可以用已映射的内存文件来共享数据。详见 MSDN（http://msdn.microsoft.com/en-us/library/windows/desktop/aa366883%28v=vs.85%29.aspx）。

正是因为 IPCDemo 在运行 Worker 进程之前就创建了文件映射，所以从 Worker 进程询问文件映射之前不用检查文件映射是否存在。IPCDemo 创建并初始化应用程序窗口和待绘制区域后，创建了一个已命名的互斥量和文件映射。然后，用不同的命令行参数（用以区别）创建不同的进程。

WndProc 例程处理 WM_COMMAND 和 WM_DESTROY 消息。当我们需要通知应用程序安全地关闭时，

WM_COMMAND 触发按钮按下事件，而 WM_DESTROY 则释放用过的文件映射，并向主线程消息队列寄送关闭消息：

```
PostQuitMessage( 0 );
```

### 更多讨论

文件映射要与常驻磁盘的文件和常驻内存的文件视图一起运作。用内存的文件视图比用硬盘驱动的读写速度快。如果要用共享对象在进程之间处理一些简单的事情，选用文件映射是很好的编程习惯。如果把 CreateFileMapping API 的第 1 个参数设置为-1，磁盘中就不会有文件存在：

```
CreateFileMapping( ( HANDLE ) -1, NULL, PAGE_READWRITE, 0,
    sizeof( COMMUNICATIONOBJECT ), COMMUNICATION_OBJECT_NAME );
```

这真是再好不过了，因为我们正打算使用一部分内存，这样更快，而且也够用了。

调用 IPCWorker 进程时要注意。像下面这样设置 CreateProcess，以供调试：

```
bSuccess = CreateProcess( TEXT( "..\\Debug\\IPCWorker.exe" ),
    TEXT( "red" ), NULL, NULL, FALSE, 0, NULL, NULL,
    &startupInfoRed, &processInformationRed );
```

Visual Studio 在调试模式中只会从项目文件夹开始启动，不会从程序的 exe 文件夹开始启动。而且，Visual Studio 默认把所有的 Win32 项目都输出到同一个文件夹中。所以，在文件路径中，我们必须从项目文件夹返回上一级（文件夹），然后找到 Debug 文件夹，整个项目的输出（exe）就在这个文件夹中。如果不想让 VS 这样启动 exe，就必须改变 CreateProcess 调用的路径，或者添加通过命令行或其他类似方法访问文件路径的功能。

## 2.6 解决典型的 IPC 问题

进程间通信非常重要，它的实现也很复杂。操作系统的设计人员和开发人员要面临各种问题。接下来，我们讲解一些最常见的问题。

### 2.6.1 哲学家就餐问题

本节讨论的哲学家就餐问题的定义，选自 Andrew S. Tanenbaum 所著的 *Mordern Operating Systems*（《现代操作系统》）第三版。作者在书中提供了解决方案。

1965 年，Dijkstra 提出并解决了一个同步问题，他称之为哲学家就餐问题。这个问题简单地描述如下：5 位哲学家围坐在一张圆桌边。每位哲学家前面都放着一盘意大利面条。面条很滑，要用两个餐叉才吃得到。相邻两个盘子之间有一个餐叉。桌上的布局如图 2.9 所示。

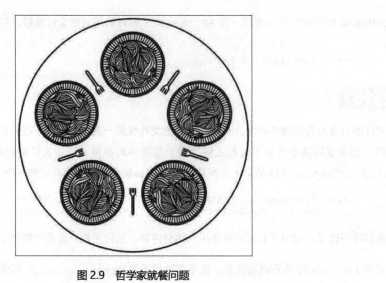

图 2.9 哲学家就餐问题

假设哲学家的状态是吃面条和思考交替进行。如果哲学家饿了，就会尝试拿起他左边和右边的餐叉，一次只能拿一把。如果成功获得两把餐叉，他就吃一会意大利面，然后放下餐叉，继续思考。关键问题是：能否写一个程序，描述每位哲学家应该怎么做才一定不会卡壳？

我们可以等指定的餐叉可用时才去拿。不过，这样想显然是错误的。如果 5 位哲学家都同时拿起左边的餐叉，就没人能拿到右边的餐叉，这就出现了死锁。

我们可以修改一下程序，在拿起左边餐叉后，程序检查右边的餐叉是否可用。如果不可用，该哲学家就放下已拿起的左边餐叉，等待一段时间，再重复这一过程。虽然这个解法和上一个解法不同，但是好不到哪里去，也是错误的。如果很不巧，所有的哲学家都同时以该算法开始，拿起他们左边的餐叉，发现右边餐叉不可用，然后放下左边餐叉，等待一会，又同时拿起左边的餐叉……这样永无止尽。这种所有程序无限期不停运行却没有任何进展的情况，叫做饥饿（*starvation*）。

要实现既不会发生死锁也不会发生饥饿，就要保护"思考"（通过互斥量调用）后面的 5 个语句。哲学家在开始拿起餐叉之前，要先询问互斥量。前面介绍过，互斥量代表相互排斥或者能给对象提供独占访问。在放下餐叉后，哲学家要释放互斥量。理论上，这种解决方案可行。但这实际上有一个性能问题：在任意时刻，只有一个哲学家进餐。桌上有 5 个餐叉可用，应该能让两个哲学家同时进餐。

下面给出了完整的解决方案。

## 准备就绪

确定安装并运行了 Visual Studio。

## 操作步骤

1. 创建一个新的默认 Win32 项目,命名为 `PhilosophersDinner`。

2. 打开 `stdafx.h`,并输入下面的代码:

   ```
   #pragma once
   #include "targetver.h"
   #define WIN32_LEAN_AND_MEAN
   #include <windows.h>
   #include <commctrl.h>
   #include <stdlib.h>
   #include <malloc.h>
   #include <memory.h>
   #include <tchar.h>
   #include <stdio.h>
   #pragma comment ( lib, "comctl32.lib" )
   #pragma comment ( linker, "\"/manifestdependency:type='win32' \
   name='Microsoft.Windows.Common-Controls' \
   version='6.0.0.0' processorArchitecture='*' \
   publicKeyToken='6595b64144ccf1df' language='*'\"" )
   ```

3. 打开 `PhilosophersDinner.cpp`,并输入下面的代码:

   ```
   #include "stdafx.h"

   #define BUTTON_CLOSE 100
   #define PHILOSOPHER_COUNT 5
   #define WM_INVALIDATE WM_USER + 1

   typedef struct _tagCOMMUNICATIONOBJECT
   {
       HWND hWnd;
       bool bExitApplication;
       int iPhilosopherArray[PHILOSOPHER_COUNT];
       int PhilosopherCount;
   } COMMUNICATIONOBJECT, *PCOMMUNICATIONOBJECT;

   HWND InitInstance(HINSTANCE hInstance, int nCmdShow);
   ATOM MyRegisterClass(HINSTANCE hInstance);
   LRESULT CALLBACK WndProc(HWND hWnd, UINT uMsg, WPARAM wParam,
       LPARAM lParam);
   int PhilosopherPass(int iPhilosopher);
   void FillEllipse(HWND hWnd, HDC hDC, int iLeft, int iTop, int
       iRight, int iBottom, int iPass);

   TCHAR* szTitle = TEXT("Philosophers Dinner Demo");
   TCHAR* szWindowClass = TEXT("__PD_WND_CLASS__");
   TCHAR* szSemaphoreName = TEXT("__PD_SEMAPHORE__");
   TCHAR* szMappingName = TEXT("__SHARED_FILE_MAPPING__");
   PCOMMUNICATIONOBJECT pCommObject = NULL;

   int APIENTRY _tWinMain(HINSTANCE hInstance, HINSTANCE
       hPrevInstance, LPTSTR lpCmdLine, int nCmdShow)
   {
       UNREFERENCED_PARAMETER(hPrevInstance);
       UNREFERENCED_PARAMETER(lpCmdLine);
   ```

```
        HANDLE hMapping = CreateFileMapping((HANDLE)-1, NULL,
            PAGE_READWRITE, 0, sizeof(COMMUNICATIONOBJECT), szMappingName);
        if (!hMapping)
        {
            MessageBox(NULL, TEXT("Cannot open file mapping"),
                TEXT("Error!"), MB_OK);
            return 1;
        }
        pCommObject = (PCOMMUNICATIONOBJECT)MapViewOfFile(hMapping,
            FILE_MAP_ALL_ACCESS, 0, 0, 0);
        if (!pCommObject)
        {
            MessageBox(NULL, TEXT("Cannot get access to file mapping! "),
                TEXT("Error!"), MB_OK);
            CloseHandle(hMapping);
            return 1;
        }
        InitCommonControls();
        MyRegisterClass(hInstance);
        HWND hWnd = NULL;
        if (!(hWnd = InitInstance(hInstance, nCmdShow)))
        {
            return FALSE;
        }
        pCommObject->bExitApplication = false;
        pCommObject->hWnd = hWnd;
        memset(pCommObject->iPhilosopherArray, 0,
            sizeof(*pCommObject->iPhilosopherArray));
        pCommObject->PhilosopherCount = PHILOSOPHER_COUNT;
        HANDLE hSemaphore = CreateSemaphore(NULL,
            int(PHILOSOPHER_COUNT / 2), int(PHILOSOPHER_COUNT / 2),
            szSemaphoreName);
        STARTUPINFO startupInfo[PHILOSOPHER_COUNT] =
            { { 0 }, { 0 }, { 0 }, { 0 }, { 0 } };
        PROCESS_INFORMATION processInformation[PHILOSOPHER_COUNT] =
            { { 0 }, { 0 }, { 0 }, { 0 }, { 0 } };
        HANDLE hProcesses[PHILOSOPHER_COUNT];
        TCHAR szBuffer[8];
        for (int iIndex = 0; iIndex < PHILOSOPHER_COUNT; iIndex++)
        {
#ifdef UNICODE
            wsprintf(szBuffer, L"%d", iIndex);
#else
            sprintf(szBuffer, "%d", iIndex);
#endif
            if (CreateProcess(TEXT("..\\Debug\\Philosopher.exe"),
                szBuffer, NULL, NULL,
                FALSE, 0, NULL, NULL, &startupInfo[iIndex],
                &processInformation[iIndex]))
            {
                hProcesses[iIndex] = processInformation[iIndex].hProcess;
            }
        }
        MSG msg = { 0 };
        while (GetMessage(&msg, NULL, 0, 0))
        {
            TranslateMessage(&msg);
            DispatchMessage(&msg);
        }
```

```
            pCommObject->bExitApplication = true;
            UnmapViewOfFile(pCommObject);
            WaitForMultipleObjects(PHILOSOPHER_COUNT, hProcesses, TRUE, INFINITE);
            for (int iIndex = 0; iIndex < PHILOSOPHER_COUNT; iIndex++)
            {
                CloseHandle(hProcesses[iIndex]);
            }
            CloseHandle(hSemaphore);
            CloseHandle(hMapping);
            return (int)msg.wParam;
        }
        ATOM MyRegisterClass(HINSTANCE hInstance)
        {
            WNDCLASSEX wndEx;
            wndEx.cbSize = sizeof(WNDCLASSEX);
            wndEx.style = CS_HREDRAW | CS_VREDRAW;
            wndEx.lpfnWndProc = WndProc;
            wndEx.cbClsExtra = 0;
            wndEx.cbWndExtra = 0;
            wndEx.hInstance = hInstance;
            wndEx.hIcon = LoadIcon(hInstance, MAKEINTRESOURCE(IDI_APPLICATION));
            wndEx.hCursor = LoadCursor(NULL, IDC_ARROW);
            wndEx.hbrBackground = (HBRUSH)(COLOR_WINDOW + 1);
            wndEx.lpszMenuName = NULL;
            wndEx.lpszClassName = szWindowClass;
            wndEx.hIconSm = LoadIcon(wndEx.hInstance, MAKEINTRESOURCE(IDI_APPLICATION));
            return RegisterClassEx(&wndEx);
        }
        HWND InitInstance(HINSTANCE hInstance, int nCmdShow)
        {
            HWND hWnd = CreateWindow(szWindowClass, szTitle, WS_OVERLAPPED
                | WS_CAPTION | WS_SYSMENU | WS_MINIMIZEBOX, 200, 200, 540, 590,
                NULL, NULL, hInstance, NULL);
            if (!hWnd)
            {
                return NULL;
            }
            HFONT hFont = CreateFont(14, 0, 0, 0, FW_NORMAL, FALSE, FALSE,
                FALSE, BALTIC_CHARSET, OUT_DEFAULT_PRECIS,
                CLIP_DEFAULT_PRECIS, DEFAULT_QUALITY, DEFAULT_PITCH |
                FF_MODERN, TEXT("Microsoft Sans Serif"));
            HWND hButton = CreateWindow(TEXT("BUTTON"), TEXT("Close"),
                WS_CHILD | WS_VISIBLE | BS_PUSHBUTTON | WS_TABSTOP, 410, 520, 100,
                25, hWnd, (HMENU)BUTTON_CLOSE, hInstance, NULL);
            SendMessage(hButton, WM_SETFONT, (WPARAM)hFont, TRUE);
            ShowWindow(hWnd, nCmdShow);
            UpdateWindow(hWnd);
            return hWnd;
        }
        LRESULT CALLBACK WndProc(HWND hWnd, UINT uMsg, WPARAM wParam, LPARAM lParam)
        {
            switch (uMsg)
            {
                case WM_COMMAND:
                {
                    switch (LOWORD(wParam))
                    {
                        case BUTTON_CLOSE:
                        {
```

```
                            DestroyWindow(hWnd);
                            break;
                    }
                    break;
            }
            case WM_INVALIDATE:
            {
                InvalidateRect(hWnd, NULL, TRUE);
                break;
            }
            case WM_PAINT:
            {
                PAINTSTRUCT paintStruct;
                HDC hDC = BeginPaint(hWnd, &paintStruct);
                FillEllipse(hWnd, hDC, 210, 10, 310, 110,
                    PhilosopherPass(1));
                FillEllipse(hWnd, hDC, 410, 170, 510, 270,
                    PhilosopherPass(2));
                FillEllipse(hWnd, hDC, 335, 400, 435, 500,
                    PhilosopherPass(3));
                FillEllipse(hWnd, hDC, 80, 400, 180, 500,
                    PhilosopherPass(4));
                FillEllipse(hWnd, hDC, 10, 170, 110, 270,
                    PhilosopherPass(5));
                EndPaint(hWnd, &paintStruct);
                break;
            }
            case WM_DESTROY:
            {
                PostQuitMessage(0);
                break;
            }
            default:
            {
                return DefWindowProc(hWnd, uMsg, wParam, lParam);
            }
        }
        return 0;
    }
    int PhilosopherPass(int iPhilosopher)
    {
        return pCommObject->iPhilosopherArray[iPhilosopher - 1];
    }
    void FillEllipse(HWND hWnd, HDC hDC, int iLeft, int iTop, int
        iRight, int iBottom, int iPass)
    {
        HBRUSH hBrush = NULL;
        if (iPass)
        {
            hBrush = CreateSolidBrush(RGB(255, 0, 0));
        }
        else
        {
            hBrush = CreateSolidBrush(RGB(255, 255, 255));
        }
        HBRUSH hOldBrush = (HBRUSH)SelectObject(hDC, hBrush);
        Ellipse(hDC, iLeft, iTop, iRight, iBottom);
        SelectObject(hDC, hOldBrush);
```

```
            DeleteObject(hBrush);
    }
```

4. 右键单击【解决方案资源管理器】,并添加一个新的默认 Win32 控制台应用程序,命名为 Philosopher。

5. 打开 stdafx.h,并输入下面的代码:

    ```
    #pragma once
    #include "targetver.h"
    #include <stdio.h>
    #include <tchar.h>
    #include <windows.h>
    ```

6. 打开 Philosopher.cpp,并输入下面的代码:

    ```
    #include "stdafx.h"
    #include <Windows.h>

    #define EATING_TIME 1000
    #define PHILOSOPHER_COUNT 5
    #define WM_INVALIDATE WM_USER + 1

    typedef struct _tagCOMMUNICATIONOBJECT
    {
        HWND hWnd;
        bool bExitApplication;
        int iPhilosopherArray[PHILOSOPHER_COUNT];
        int PhilosopherCount;
    } COMMUNICATIONOBJECT, *PCOMMUNICATIONOBJECT;

    void Eat();
    TCHAR* szSemaphoreName = TEXT("__PD_SEMAPHORE__");
    TCHAR* szMappingName = TEXT("__SHARED_FILE_MAPPING__");
    bool bExitApplication = false;

    int _tmain(int argc, _TCHAR* argv[])
    {
        HWND hConsole = GetConsoleWindow();
        ShowWindow(hConsole, SW_HIDE);
        int iIndex = (int)_tcstol(argv[0], NULL, 10);
        HANDLE hMapping = OpenFileMapping(FILE_MAP_ALL_ACCESS, FALSE,
            szMappingName);
        while (!bExitApplication)
        {
            HANDLE hSemaphore = OpenSemaphore(SEMAPHORE_ALL_ACCESS, FALSE,
                szSemaphoreName);
            WaitForSingleObject(hSemaphore, INFINITE);
            PCOMMUNICATIONOBJECT pCommObject = (PCOMMUNICATIONOBJECT)
                MapViewOfFile(hMapping, FILE_MAP_ALL_ACCESS, 0, 0,
                    sizeof(COMMUNICATIONOBJECT));
            bExitApplication = pCommObject->bExitApplication;
            if (!pCommObject->iPhilosopherArray[
                (iIndex + pCommObject->PhilosopherCount - 1)
                    % pCommObject->PhilosopherCount]
                    && !pCommObject->iPhilosopherArray[
                        (iIndex + 1) % pCommObject->PhilosopherCount])
            {
    ```

```
                pCommObject->iPhilosopherArray[iIndex] = 1;
                Eat();
            }

            SendMessage(pCommObject->hWnd, WM_INVALIDATE, 0, 0);
            pCommObject->iPhilosopherArray[iIndex] = 0;
            UnmapViewOfFile(pCommObject);
            ReleaseSemaphore(hSemaphore, 1, NULL);
            CloseHandle(hSemaphore);
        }
        CloseHandle(hMapping);
        return 0;
    }

    void Eat()
    {
        Sleep(EATING_TIME);
    }
```

## 示例分析

我们要创建 5 个进程来模仿 5 位哲学家的行为。每位哲学家（即，每个进程）都必须思考和进餐。哲学家需要两把餐叉才能进餐，在餐叉可用的前提下，他必须先拿起左边的餐叉，再拿起右边的餐叉。如果两把餐叉都可用，他就能顺利进餐；如果另一把餐叉不可用，他就放下已拿起的左边餐叉，并等待下一次进餐。我们在程序中把进餐时间设置为 1 秒。

PhilosophersDinner 是该主应用程序。我们创建了文件映射，可以与其他进程通信。创建 Semaphore 对象同步进程也很重要。根据前面的分析，使用互斥量能确保同一时刻只有一位哲学家进餐。这种虽然方法可行，但是优化得不够。如果每位哲学家需要两把餐叉才能进餐，可以用 FLOOR( NUMBER_OF_PHILOSOPHERS / 2 )实现两位哲学家同时进餐。这就是我们设置同一时刻最多有两个对象可以传递信号量的原因，如下代码所示：

```
HANDLE hSemaphore = CreateSemaphore( NULL,
    int( PHILOSOPHER_COUNT / 2 ),
    int( PHILOSOPHER_COUNT / 2 ), szSemaphoreName );
```

这里注意，信号量最初可以允许一定数量的对象通过，但是通过 CreateSemaphore API 的第 3 个参数可以递增这个数量。不过在我们的示例中，用不到这个特性。

初始化信号量对象后，就创建了进程，应用程序可以进入消息循环了。分析完 PhilosophersDinner，我们来看 Philosopher 应用程序。这是一个控制台应用程序，因为我们不需要接口，所以将隐藏它的主窗口（本例中是控制台）。如下代码所示：

```
HWND hConsole = GetConsoleWindow( );
ShowWindow( hConsole, SW_HIDE );
```

接下来，该应用程序要获得它的索引（哲学家的姓名）：

```
int iIndex = ( int ) _tcstol( argv[ 0 ], NULL, 10 );
```

然后，哲学家必须获得文件映射对象的句柄，并进入消息循环。在消息循环中，哲学家询问传递过来的信号量对象，等待轮到他们进餐。当哲学家获得一个传入的信号量时，就可以获得两把餐叉。然后，通过下面的 SendMessage API，发送一条消息，更新主应用程序的用户接口：

```
SendMessage( pCommObject->hWnd, WM_INVALIDATE, 0, 0 );
```

所有的工作完成后，哲学家会释放信号量对象并继续执行。

### 更多讨论

还有一些经典的 IPC 问题，如"睡觉的理发师"和"生产者-消费者"问题。本章后面会给出"生产者-消费者"问题的解法。

## 2.7　线程模型的实现

我们可以把进程看作是一个对象，它的任务就是把相关资源分组。每个进程都有一个地址空间，如图 2.10 所示。

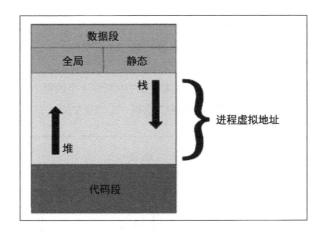

图 2.10　进程的地址空间

这个所谓的进程图像必须在初始化 CreateProcess 时加载至物理内存中。所有的资源（如文件句柄、子进程的信息、信号处理器等）都被储存起来。把它们以进程的形式分组在一起，更容易管理。

除进程外，还有一个重要的概念是**线程**。线程是 CPU 可执行调度的最小单位。也就是说，进程本身不能获得 CPU 时间，只有它的线程才可以。线程通过它的工作变量和栈来储存 CPU 寄存器的信息。栈包含与函数调用相关的数据，在每个函数被调用但尚未返回时，为其创建一个框架。线程可以在 CPU 上执行，而进程则不行。但是，进程至少必须有一个线程，通常把这个线程称为主线程。因此，当我们说在 CPU 上执行的进程时，指的是进程中的主线程。

进程用于分组资源，线程是在 CPU 上调度执行的实体。在同一个进程环境中可以执行多个线程，理解这点很重要。多线程并行运行在一个进程上下文，与在一个计算机中并行运行的多个进程相同。术语"多线程"指的是在单进程上下文中运行的多线程。

如图 2.11 所示，有 3 个进程，每个进程中都有一个线程。

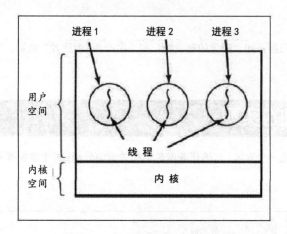

图 2.11　3 个进程中各有 1 个线程

图 2.12 演示了一个有 3 个线程的进程。虽然这两种情况中都有 3 个线程，但是在图 2.11 中，每个线程都在不同的地址空间中运行，而图 2.12 中的 3 个线程共享同一个地址空间。

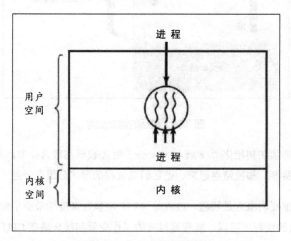

图 2.12　有 3 个线程的进程

## 2.7 线程模型的实现

在单核 CPU 系统中运行多线程的进程时,各线程轮流运行。系统通过快速切换多个进程,营造并行处理的假象。多线程也以这样的方式运行。一个有 3 个线程的进程,其各线程表现为并行运行。单核 CPU 每次运行一个线程,花费 CUP 调度处理该进程时间的 1/3(大概是这样,CPU 时间取决于操作系统、调度算法等)。在多处理器系统中,情况类似。只有单核 CPU 执行线程时才与本书描述的方式相同。多核的好处是,可以并行运行更多的线程,充分发挥本地硬件的并行处理能力和多线程的执行能力。

下面的例子用两个线程实现一个简单的数组排序,演示了线程的基本用法。

### 准备就绪

确定安装并运行了 Visual Studio。

### 操作步骤

1. 创建一个新的默认 Win32 控制台应用程序,名为 MultithreadedArraySort。

2. 打开 MultithreadedArraySort.cpp,并输入下面的代码:

```cpp
#include "stdafx.h"
#include <Windows.h>
#include <iostream>
#include <tchar.h>

using namespace std;

#define THREADS_NUMBER 2
#define ELEMENTS_NUMBER 200
#define BLOCK_SIZE ELEMENTS_NUMBER / THREADS_NUMBER
#define MAX_VALUE 1000

typedef struct _tagARRAYOBJECT
{
    int* iArray;
    int iSize;
    int iThreadID;
} ARRAYOBJECT, *PARRAYOBJECT;

DWORD WINAPI ThreadStart(LPVOID lpParameter);
void PrintArray(int* iArray, int iSize);
void MergeArrays(int* leftArray, int leftArrayLenght, int*
    rightArray, int rightArrayLenght, int* mergedArray);

int _tmain(int argc, TCHAR* argv[])
{
    int iArray1[BLOCK_SIZE];
    int iArray2[BLOCK_SIZE];
    int iArray[ELEMENTS_NUMBER];
    for (int iIndex = 0; iIndex < BLOCK_SIZE; iIndex++)
    {
        iArray1[iIndex] = rand() % MAX_VALUE;
        iArray2[iIndex] = rand() % MAX_VALUE;
    }
```

```cpp
        HANDLE hThreads[THREADS_NUMBER];
        ARRAYOBJECT pObject1 = { &(iArray1[0]), BLOCK_SIZE, 0 };
        hThreads[0] = CreateThread(NULL, 0, (LPTHREAD_START_ROUTINE)
            ThreadStart, (LPVOID)&pObject1, 0, NULL);

        ARRAYOBJECT pObject2 = { &(iArray2[0]), BLOCK_SIZE, 1 };
        hThreads[1] = CreateThread(NULL, 0, (LPTHREAD_START_ROUTINE)
            ThreadStart, (LPVOID)&pObject2, 0, NULL);

        cout << "Waiting execution..." << endl;
        WaitForMultipleObjects(THREADS_NUMBER, hThreads, TRUE, INFINITE);

        MergeArrays(&iArray1[0], BLOCK_SIZE, &iArray2[0], BLOCK_SIZE, &iArray[0]);
        PrintArray(iArray, ELEMENTS_NUMBER);

        CloseHandle(hThreads[0]);
        CloseHandle(hThreads[1]);

        cout << "Array sorted..." << endl;
        return 0;
    }

    DWORD WINAPI ThreadStart(LPVOID lpParameter)
    {
        PARRAYOBJECT pObject = (PARRAYOBJECT)lpParameter;
        int iTmp = 0;
        for (int iIndex = 0; iIndex < pObject->iSize; iIndex++)
        {
            for (int iEndIndex = pObject->iSize - 1; iEndIndex > iIndex; iEndIndex--)
            {
                if (pObject->iArray[iEndIndex] < pObject->iArray[iIndex])
                {
                    iTmp = pObject->iArray[iEndIndex];
                    pObject->iArray[iEndIndex] = pObject->iArray[iIndex];
                    pObject->iArray[iIndex] = iTmp;
                }
            }
        }
        return 0;
    }

    void PrintArray(int* iArray, int iSize)
    {
        for (int iIndex = 0; iIndex < iSize; iIndex++)
        {
            cout << " " << iArray[iIndex];
        }
        cout << endl;
    }

    void MergeArrays(int* leftArray, int leftArrayLenght, int*
        rightArray, int rightArrayLenght, int* mergedArray)
    {
        int i = 0;
        int j = 0;
        int k = 0;
        while (i < leftArrayLenght && j < rightArrayLenght)
        {
```

```
            if (leftArray[i] < rightArray[j])
            {
                mergedArray[k] = leftArray[i];
                i++;
            }
            else
            {
                mergedArray[k] = rightArray[j];
                j++;
            }
            k++;
        }
        if (i >= leftArrayLenght)
        {
            while (j < rightArrayLenght)
            {
                mergedArray[k] = rightArray[j];
                j++;
                k++;
            }
        }
        if (j >= rightArrayLenght)
        {
            while (i < leftArrayLenght)
            {
                mergedArray[k] = leftArray[i];
                i++;
                k++;
            }
        }
    }
```

## 示例分析

这个程序示例很简单,演示了线程的基本用法。该示例背后的思想是,为了节省执行时间而添加并行,把问题划分为几个小问题,并分配给几个线程(分而治之)。我们在前面提到过,把问题划分成若干更小的单元,更容易在实现中创建并行逻辑。同时,在并行中使用系统资源能优化应用程序并提高其运行速度。

## 更多讨论

如前所述,每个应用程序都有一个主线程。使用 CreateThread Win32 API 创建其他线程:

```
HANDLE CreateThread( LPSECURITY_ATTRIBUTES lpThreadAttributes,
    SIZE_T dwStackSize, LPTHREAD_START_ROUTINE lpStartAddress,
    LPVOID lpParameter, DWORD dwFlags, LPDWORD lpThreadId );
```

设置线程的开始地址(lpStartAddress)和设置传给线程例程的值(lpParameter)很重要。lpParameter 是一个预定义例程(函数)指针,如下代码所示:

```
typedef DWORD ( WINAPI *PTHREAD_START_ROUTINE )( LPVOID lpThreadParameter );
```

我们的 ThreadStart 方法与指定的原型匹配,这也是开始执行线程的地方。CreateThread API 的第 4 个参数是一个要传递给线程例程的指针。如果要传递更多参数,可以创建一个结构或类,然后传递相应对象的地址。欲详细了解 CreateThread API，请查阅 MSDN（http://msdn.microsoft.com/en-us/library/windows/desktop/ ms682453%28v=vs.85%29.aspx）。

## 2.8　线程的用法

现在大部分应用程序都使用一些数据库。在许多情况下，这种应用程序通常会运行在不同的 PC 中，并同时进行读写操作。下面例子中的线程使用了 MySQL 数据库。

### 准备就绪

该示例要求安装 MySQL C Connector，详情请查阅附录。成功安装 MySQL C Connector 后，运行 Visual Studio。

### 操作步骤

1. 创建一个新的默认 C++控制台应用程序，命名为 MultithreadedDBTest。

2. 打开【解决方案资源管理器】，添加一个新的头文件，命名为 CMYSQL.h。打开 CMYSQL.h，并输入下面的代码：

```cpp
#include "stdafx.h"
#include <stdio.h>
#include <stdlib.h>
#include <mysql.h>

class CMySQL
{
public:
    static CMySQL* CreateInstance(char* szHostName, char* szDatabase,
        char* szUserId, char* szPassword);
    static void ReleaseInstance();
    bool ConnectInstance();
    bool DisconnectInstance();
    bool ReadData(char* szQuery, char* szResult, size_t uBufferLenght);
    bool WriteData(char* szQuery, char* szResult, size_t uBufferLenght);
private:
    CMySQL(char* szHostName, char* szDatabase, char* szUserId, char*
        szPassword);
    ~CMySQL();
    char* szHostName;
    char* szDatabase;
    char* szUserId;
    char* szPassword;
    MYSQL* mysqlConnection;
    static CMySQL* mySqlInstance;
};
```

3. 打开【解决方案资源管理器】，添加一个新的 CMYSQL.cpp 文件。打开 CMYSQL.cpp，并输入下

面的代码:

```cpp
#include "stdafx.h"
#include "CMySQL.h"

CMySQL* CMySQL::mySqlInstance = NULL;
CMySQL* CMySQL::CreateInstance(char* szHostName, char* szDatabase,
    char* szUserId, char* szPassword)
{
    if (mySqlInstance)
    {
        return mySqlInstance;
    }
    return new CMySQL(szHostName, szDatabase, szUserId, szPassword);
}
void CMySQL::ReleaseInstance()
{
    if (mySqlInstance)
    {
        delete mySqlInstance;
    }
}

CMySQL::CMySQL(char* szHostName, char* szDatabase, char* szUserId,
   char* szPassword)
{
    size_t length = 0;
    this->szHostName = new char[length = strlen(szHostName) + 1];
    strcpy_s(this->szHostName, length, szHostName);
    this->szDatabase = new char[length = strlen(szDatabase) + 1];
    strcpy_s(this->szDatabase, length, szDatabase);
    this->szUserId = new char[length = strlen(szUserId) + 1];
    strcpy_s(this->szUserId, length, szUserId);

    this->szPassword = new char[length = strlen(szPassword) + 1];
    strcpy_s(this->szPassword, length, szPassword);
}

CMySQL::~CMySQL()
{
    delete szHostName;
    delete szDatabase;
    delete szUserId;
    delete szPassword;
}

bool CMySQL::ConnectInstance()
{
    MYSQL* mysqlLink = NULL;

    try
    {
        mysqlConnection = mysql_init(NULL);
        mysqlLink = mysql_real_connect(mysqlConnection, szHostName,
            szUserId, szPassword, szDatabase, 3306, NULL, 0);
    }
    catch (...)
    {
        mysqlConnection = 0;
```

```cpp
            return false;
    }
    return mysqlLink ? true : false;
}

bool CMySQL::DisconnectInstance()
{
    try
    {
        mysql_close(mysqlConnection);
        return true;
    }
    catch (...)
    {
        return false;
    }
}

bool CMySQL::ReadData(char* szQuery, char* szResult, size_t uBufferLenght)
{
    int mysqlStatus = 0;
    MYSQL_RES* mysqlResult = NULL;
    MYSQL_ROW mysqlRow = NULL;
    my_ulonglong numRows = 0;
    unsigned numFields = 0;
    try
    {
        mysqlStatus = mysql_query(mysqlConnection, szQuery);
        if (mysqlStatus)
        {
            return false;
        }
        else
        {
            mysqlResult = mysql_store_result(mysqlConnection);
        }
        if (mysqlResult)
        {
            numRows = mysql_num_rows(mysqlResult);
            numFields = mysql_num_fields(mysqlResult);
        }
        mysqlRow = mysql_fetch_row(mysqlResult);
        if (mysqlRow)
        {
            if (!mysqlRow[0])
            {
                mysql_free_result(mysqlResult);
                return false;
            }
        }
        else
        {
            mysql_free_result(mysqlResult);
            return false;
        }

        size_t szResultLength = strlen(mysqlRow[0]) + 1;
        strcpy_s(szResult, szResultLength > uBufferLenght ?
            uBufferLenght : szResultLength, mysqlRow[0]);
```

```cpp
            if (mysqlResult)
            {
                mysql_free_result(mysqlResult);
                mysqlResult = NULL;
            }
        }
        catch (...)
        {
            return false;
        }
        return true;
    }

    bool CMySQL::WriteData(char* szQuery, char* szResult, size_t
        uBufferLenght)
    {
        try
        {
            int mysqlStatus = mysql_query(mysqlConnection, szQuery);
            if (mysqlStatus)
            {
                size_t szResultLength = strlen("Failed!") + 1;
                strcpy_s(szResult, szResultLength > uBufferLenght ?
                    uBufferLenght : szResultLength, "Failed!");
                return false;
            }
        }
        catch (...)
        {
            size_t szResultLength = strlen("Exception!") + 1;
            strcpy_s(szResult, szResultLength > uBufferLenght ?
                uBufferLenght : szResultLength, "Exception!");
            return false;
        }

        size_t szResultLength = strlen("Success") + 1;
        strcpy_s(szResult, szResultLength > uBufferLenght ?
            uBufferLenght : szResultLength, "Success");
        return true;
    }
```

4. 打开 MultithreadedDBTest.cpp，并添加下面的代码：

```cpp
#include "stdafx.h"
#include "CMySQL.h"

#define BLOCK_SIZE 4096
#define THREADS_NUMBER 3

typedef struct
{
    char szQuery[BLOCK_SIZE];
    char szResult[BLOCK_SIZE];
    bool bIsRead;
} QUERYDATA, *PQUERYDATA;

CRITICAL_SECTION cs;
CMySQL* mySqlInstance = NULL;
```

```c
DWORD WINAPI StartAddress(LPVOID lpParameter)
{
    PQUERYDATA pQueryData = (PQUERYDATA)lpParameter;
    EnterCriticalSection(&cs);
    if (mySqlInstance->ConnectInstance())
    {
        if (pQueryData->bIsRead)
        {
            memset(pQueryData->szResult, 0, BLOCK_SIZE - 1);
            mySqlInstance->ReadData(pQueryData->szQuery,
                pQueryData->szResult, BLOCK_SIZE - 1);
        }
        else
        {
            mySqlInstance->WriteData(pQueryData->szQuery,
                pQueryData->szResult, BLOCK_SIZE - 1);
        }
        mySqlInstance->DisconnectInstance();
    }
    LeaveCriticalSection(&cs);
    return 0L;
}

int main()
{
    InitializeCriticalSection(&cs);
    mySqlInstance = CMySQL::CreateInstance(
        "mysql.services.expert.its.me", "expertit_9790OS",
        "expertit_9790", "$dbpass_1342#");
    if (mySqlInstance)
    {
        HANDLE hThreads[THREADS_NUMBER];
        QUERYDATA queryData[THREADS_NUMBER] =
        {
            { "select address from clients where id = 3;",
            "", true },
            { "update clients set name='Merrill & Lynch' where id=2;",
            "", false },
            { "select name from clients where id = 2;",
            "", true }
        };
        for (int iIndex = 0; iIndex < THREADS_NUMBER; iIndex++)
        {
            hThreads[iIndex] = CreateThread(NULL, 0,
                (LPTHREAD_START_ROUTINE)StartAddress,
                &queryData[iIndex], 0, 0);
        }
        WaitForMultipleObjects(THREADS_NUMBER, hThreads, TRUE, INFINITE);
        for (int iIndex = 0; iIndex < THREADS_NUMBER; iIndex++)
        {
            printf_s("%s\n", queryData[iIndex].szResult);
        }
        CMySQL::ReleaseInstance();
    }
    DeleteCriticalSection(&cs);
    return system("pause");
}
```

## 示例分析

上面的示例演示了在操作 MySQL 数据库的应用程序中实现线程同步。该例只使用了一种方法完成同步，另一种使用线程同步的方法是，通过**数据库管理系统**（Database Management System，DBMS）本身的机制，使用表锁（table lock）。这里的重点是，两个线程不能在同一时间内同时执行读或写操作，一个线程执行了读/写操作，另一个线程就不能这样做。我们假设 3 个线程中有 2 个线程必须进行读操作，而第 3 个线程必须写入某些数据。还可以假设线程在读取之后，必须给主应用程序或某些其他处理程序发读取数据的信号，然后继续处理。写入数据的线程也是这样，从其他处理程序获得数据后要执行必要的操作（发信号）。本例中我们为了同步线程，使用了一个临界区对象。临界区对象将在第 3 章详细介绍。

## 更多讨论

上面提到，在操作 MySQL 数据库时还有一种更方便的方法同步线程。对于现代的 MySQL DBMS，可以用表锁来完成。这里只是给读者提供一个思路，我们将演示一种方法，两个线程执行查询的同时还保持同步的读写操作，互不干扰。

1. 读线程：

    表锁 TABLE_NAME 读；
    在 TABLE_NAME 中选择；
    解除表锁。

2. 写线程：

    表锁 TABLE_NAME 读；
    在 TABLE_NAME 中插入值（…）；
    更新 TABLE_NAME 设置；
    解除表锁。

3. 单线程中的操作——其他线程必须等待：

    表锁 TABLE_NAME TABLE_ALIAS 读，TABLE_NAME 写；
    在 TABLE_NAME 中选择一些内容作为 TABLE_ALIAS；
    在 TABLE_NAME 中插入值…；
    解除表锁。

注意，MySQL 有多种引擎（InnoDB、MyISAM、Memory 等），不同的引擎其表锁和锁粒度都不同。

## 2.9 在用户空间实现线程

既可以在用户空间也可以在内核中实现线程包。具体选择在哪里实现还存在一些争议，在一些实现中可能会混合使用内核线程和用户线程。

我们将讨论在不同地方实现线程包的方法及优缺点。第 1 种方法是，把整个线程包放进用户空间，内核完全不知道。就内核而言，它管理着普通的单线程进程。这种方法的优点和最显著的优势是，可以在不支持线程的操作系统中实现用户级线程包。

过去，传统的操作系统就采用这种方法，甚至沿用至今。用这种方法，线程可以通过库来实现。所有这些实现都具有相同的通用结构。线程运行在运行时系统的顶部，该系统是专门管理线程的过程集合。我们在前面见过一些例子（`CreateThread`、`TerminateThread` 等），以后还会见到更多。

下面的程序示例演示了在用户空间中的线程用法。我们要复制大型文件，但是不想一开始就读取整个文件的内容，或者更优化地一部分一部分地读取，而且不用在文件中写入数据。这就涉及 2.5 节中提到的**生产者-消费者**问题。

### 准备就绪

确定安装并运行了 Visual Studio。

### 操作步骤

1. 创建一个新的 Win32 应用程序项目，并命名为 `ConcurrentFileCopy`。

2. 打开【解决方案资源管理器】，添加一个新的头文件，命名为 `ConcurrentFileCopy.h`。打开 `ConcurrentFileCopy.h`，并输入下面的代码：

   ```
   #pragma once

   #include <windows.h>
   #include <commctrl.h>
   #include <memory.h>
   #include <tchar.h>
   #include <math.h>

   #pragma comment ( lib, "comctl32.lib" )
   #pragma comment ( linker, "\"/manifestdependency:type='win32' \
   name='Microsoft.Windows.Common-Controls' \
   version='6.0.0.0' processorArchitecture='*' \
   publicKeyToken='6595b64144ccf1df' language='*'\"" )

   ATOM RegisterWndClass(HINSTANCE hInstance);
   HWND InitializeInstance(HINSTANCE hInstance, int nCmdShow, HWND& hWndPB);
   LRESULT CALLBACK WndProc(HWND hWnd, UINT uMsg, WPARAM wParam, LPARAM lParam);
   DWORD WINAPI ReadRoutine(LPVOID lpParameter);
   ```

```
DWORD WINAPI WriteRoutine(LPVOID lpParameter);
BOOL FileDialog(HWND hWnd, LPTSTR szFileName, DWORD
    dwFileOperation);
DWORD GetBlockSize(DWORD dwFileSize);

#define BUTTON_CLOSE 100
#define FILE_SAVE 0x0001
#define FILE_OPEN 0x0002
#define MUTEX_NAME _T("__RW_MUTEX__")

typedef struct _tagCOPYDETAILS
{
    HINSTANCE hInstance;
    HWND hWndPB;
    LPTSTR szReadFileName;
    LPTSTR szWriteFileName;
} COPYDETAILS, *PCOPYDETAILS;
```

3. 现在,打开【解决方案资源管理器】,并添加一个新的源文件,命名为ConcurrentFileCopy.cpp。
打开ConcurrentFileCopy.c,并输入下面的代码:

```
#include "ConcurrentFileCopy.h"

TCHAR* szTitle = _T("Concurrent file copy");
TCHAR* szWindowClass = _T("__CFC_WND_CLASS__");

DWORD dwReadBytes = 0;
DWORD dwWriteBytes = 0;

DWORD dwBlockSize = 0;
DWORD dwFileSize = 0;

HLOCAL pMemory = NULL;

int WINAPI _tWinMain(HINSTANCE hInstance, HINSTANCE hPrev, LPTSTR
    szCmdLine, int iCmdShow)
{
    UNREFERENCED_PARAMETER(hPrev);
    UNREFERENCED_PARAMETER(szCmdLine);

    RegisterWndClass(hInstance);

    HWND hWnd = NULL;
    HWND hWndPB = NULL;

    if (!(hWnd = InitializeInstance(hInstance, iCmdShow, hWndPB)))
    {
        return 1;
    }

    MSG msg = { 0 };
    TCHAR szReadFile[MAX_PATH];
    TCHAR szWriteFile[MAX_PATH];

    if (FileDialog(hWnd, szReadFile, FILE_OPEN) && FileDialog(hWnd,
        szWriteFile, FILE_SAVE))
    {
        COPYDETAILS copyDetails = { hInstance, hWndPB, szReadFile,
```

```
                szWriteFile };
            HANDLE hMutex = CreateMutex(NULL, FALSE, MUTEX_NAME);
            HANDLE hReadThread = CreateThread(NULL, 0,
                (LPTHREAD_START_ROUTINE)ReadRoutine, &copyDetails, 0, NULL);
            while (GetMessage(&msg, NULL, 0, 0))
            {
                TranslateMessage(&msg);
                DispatchMessage(&msg);
            }
            CloseHandle(hReadThread);
            CloseHandle(hMutex);
        }
        else
        {
            MessageBox(hWnd, _T("Cannot open file!"),
                _T("Error!"), MB_OK);
        }
        LocalFree(pMemory);
        UnregisterClass(szWindowClass, hInstance);
        return (int)msg.wParam;
}

ATOM RegisterWndClass(HINSTANCE hInstance)
{
    WNDCLASSEX wndEx;

    wndEx.cbSize = sizeof(WNDCLASSEX);
    wndEx.style = CS_HREDRAW | CS_VREDRAW;
    wndEx.lpfnWndProc = WndProc;
    wndEx.cbClsExtra = 0;
    wndEx.cbWndExtra = 0;
    wndEx.hInstance = hInstance;
    wndEx.hIcon = LoadIcon(hInstance, MAKEINTRESOURCE(IDI_APPLICATION));
    wndEx.hCursor = LoadCursor(NULL, IDC_ARROW);
    wndEx.hbrBackground = (HBRUSH)(COLOR_WINDOW + 1);
    wndEx.lpszMenuName = NULL;
    wndEx.lpszClassName = szWindowClass;
    wndEx.hIconSm = LoadIcon(wndEx.hInstance, MAKEINTRESOURCE(IDI_APPLICATION));

    return RegisterClassEx(&wndEx);
}

HWND InitializeInstance(HINSTANCE hInstance, int iCmdShow, HWND& hWndPB)
{
    HWND hWnd = CreateWindow(szWindowClass, szTitle, WS_OVERLAPPED
        | WS_CAPTION | WS_SYSMENU | WS_MINIMIZEBOX, 200, 200, 440, 290,
        NULL, NULL, hInstance, NULL);
    RECT rcClient = { 0 };
    int cyVScroll = 0;

    if (!hWnd)
    {
        return NULL;
    }

    HFONT hFont = CreateFont(14, 0, 0, 0, FW_NORMAL, FALSE, FALSE,
        FALSE, BALTIC_CHARSET, OUT_DEFAULT_PRECIS, CLIP_DEFAULT_PRECIS,
        DEFAULT_QUALITY, DEFAULT_PITCH | FF_MODERN,
        _T("Microsoft Sans Serif"));
```

```c
        HWND hButton = CreateWindow(_T("BUTTON"), _T("Close"), WS_CHILD
            | WS_VISIBLE | BS_PUSHBUTTON | WS_TABSTOP, 310, 200, 100, 25,
            hWnd, (HMENU)BUTTON_CLOSE, hInstance, NULL);
        SendMessage(hButton, WM_SETFONT, (WPARAM)hFont, TRUE);

        GetClientRect(hWnd, &rcClient);
        cyVScroll = GetSystemMetrics(SM_CYVSCROLL);

        hWndPB = CreateWindow(PROGRESS_CLASS, (LPTSTR)NULL, WS_CHILD |
            WS_VISIBLE, rcClient.left, rcClient.bottom - cyVScroll,
            rcClient.right, cyVScroll, hWnd, (HMENU)0, hInstance, NULL);
        SendMessage(hWndPB, PBM_SETSTEP, (WPARAM)1, 0);

        ShowWindow(hWnd, iCmdShow);
        UpdateWindow(hWnd);

        return hWnd;
}

LRESULT CALLBACK WndProc(HWND hWnd, UINT uMsg, WPARAM wParam, LPARAM lParam)
{
        switch (uMsg)
        {
            case WM_COMMAND:
            {
                switch (LOWORD(wParam))
                {
                    case BUTTON_CLOSE:
                    {
                        DestroyWindow(hWnd);
                        break;
                    }
                }
                break;
            }
            case WM_DESTROY:
            {
                PostQuitMessage(0);
                break;
            }
            default:
            {
                return DefWindowProc(hWnd, uMsg, wParam, lParam);
            }
        }
        return 0;
}

DWORD WINAPI ReadRoutine(LPVOID lpParameter)
{
        PCOPYDETAILS pCopyDetails = (PCOPYDETAILS)lpParameter;
        HANDLE hFile = CreateFile(pCopyDetails->szReadFileName,
            GENERIC_READ, FILE_SHARE_READ, NULL, OPEN_EXISTING,
            FILE_ATTRIBUTE_NORMAL, NULL);
        if (hFile == (HANDLE)INVALID_HANDLE_VALUE)
        {
            return FALSE;
        }
        dwFileSize = GetFileSize(hFile, NULL);
```

```
        dwBlockSize = GetBlockSize(dwFileSize);
        HANDLE hWriteThread = CreateThread(NULL, 0,
            (LPTHREAD_START_ROUTINE)WriteRoutine, pCopyDetails, 0, NULL);
        size_t uBufferLength = (size_t)ceil((double)dwFileSize/(double)dwBlockSize);
        SendMessage(pCopyDetails->hWndPB, PBM_SETRANGE, 0,
            MAKELPARAM(0, uBufferLength));
        pMemory = LocalAlloc(LPTR, dwFileSize);
        void* pBuffer = LocalAlloc(LPTR, dwBlockSize);

        int iOffset = 0;
        DWORD dwBytesRed = 0;

        do
        {
            ReadFile(hFile, pBuffer, dwBlockSize, &dwBytesRed, NULL);
            if (!dwBytesRed)
            {
                break;
            }
            HANDLE hMutex = OpenMutex(MUTEX_ALL_ACCESS, FALSE,
                MUTEX_NAME);
            WaitForSingleObject(hMutex, INFINITE);
            memcpy((char*)pMemory + iOffset, pBuffer, dwBytesRed);
            dwReadBytes += dwBytesRed;

            ReleaseMutex(hMutex);

            iOffset += (int)dwBlockSize;
        } while (true);

        LocalFree(pBuffer);
        CloseHandle(hFile);
        CloseHandle(hWriteThread);
        return 0;
    }

    DWORD WINAPI WriteRoutine(LPVOID lpParameter)
    {
        PCOPYDETAILS pCopyDetails = (PCOPYDETAILS)lpParameter;
        HANDLE hFile = CreateFile(pCopyDetails->szWriteFileName,
            GENERIC_WRITE, 0, NULL, CREATE_ALWAYS, FILE_ATTRIBUTE_NORMAL, NULL);
        if (hFile == (HANDLE)INVALID_HANDLE_VALUE)
        {
            return FALSE;
        }

        DWORD dwBytesWritten = 0;
        int iOffset = 0;

        do
        {
            int iRemainingBytes = (int)dwFileSize - iOffset;
            if (iRemainingBytes <= 0)
            {
                break;
            }
            Sleep(10);
            if (dwWriteBytes < dwReadBytes)
            {
```

```
                    DWORD dwBytesToWrite = dwBlockSize;
                    if (!(dwFileSize / dwBlockSize))
                    {
                        dwBytesToWrite = (DWORD)iRemainingBytes;
                    }
                    HANDLE hMutex = OpenMutex(MUTEX_ALL_ACCESS, FALSE, MUTEX_NAME);
                    WaitForSingleObject(hMutex, INFINITE);

                    WriteFile(hFile, (char*)pMemory + iOffset, dwBytesToWrite,
                        &dwBytesWritten, NULL);
                    dwWriteBytes += dwBytesWritten;

                    ReleaseMutex(hMutex);

                    SendMessage(pCopyDetails->hWndPB, PBM_STEPIT, 0, 0);
                    iOffset += (int)dwBlockSize;
            }
    } while (true);

    CloseHandle(hFile);
    return 0;
}

BOOL FileDialog(HWND hWnd, LPTSTR szFileName, DWORD dwFileOperation)
{
#ifdef _UNICODE
    OPENFILENAMEW ofn;
#else
    OPENFILENAMEA ofn;
#endif
    TCHAR szFile[MAX_PATH];

    ZeroMemory(&ofn, sizeof(ofn));
    ofn.lStructSize = sizeof(ofn);
    ofn.hwndOwner = hWnd;
    ofn.lpstrFile = szFile;
    ofn.lpstrFile[0] = '\0';
    ofn.nMaxFile = sizeof(szFile);
    ofn.lpstrFilter = _T("All\0*.*\0Text\0*.TXT\0");
    ofn.nFilterIndex = 1;
    ofn.lpstrFileTitle = NULL;
    ofn.nMaxFileTitle = 0;
    ofn.lpstrInitialDir = NULL;
    ofn.Flags = dwFileOperation == FILE_OPEN ? OFN_PATHMUSTEXIST |
        OFN_FILEMUSTEXIST : OFN_SHOWHELP | OFN_OVERWRITEPROMPT;

    if (dwFileOperation == FILE_OPEN)
    {
        if (GetOpenFileName(&ofn) == TRUE)
        {
            _tcscpy_s(szFileName, MAX_PATH - 1, szFile);
            return TRUE;
        }
    }
    else
    {
        if (GetSaveFileName(&ofn) == TRUE)
        {
            _tcscpy_s(szFileName, MAX_PATH - 1, szFile);
```

```
                return TRUE;
            }
        }
        return FALSE;
    }
    DWORD GetBlockSize(DWORD dwFileSize)
    {
        return dwFileSize > 4096 ? 4096 : 512;
    }
```

## 示例分析

我们创建了一个和哲学家就餐示例非常像的 UI。例程 `MyRegisterClass`、`InitInstance` 和 `WndProc` 几乎都一样。我们在程序中添加 `FileDialog` 来询问用户读写文件的路径。为了读和写，分别启动了两个线程。

操作系统的调度十分复杂。我们根本不知道是调度算法还是硬件中断使得某线程被调度在 CUP 中执行。这意味着写线程可能在读线程之前执行。出现这种情况会导致一个异常，因为写线程没东西可写。

因此，我们在写操作中添加了 `if` 条件，如下代码所示：

```
if ( dwBytesWritten < dwBytesRead )
{
   WriteFile(hFile, pCharTmp, sizeof(TCHAR) * BLOCK_SIZE, &dwFileSize, NULL);
   dwBytesWritten += dwFileSize;
   SendMessage( hProgress, PBM_STEPIT, 0, 0 );
}
```

线程在获得互斥量后，才能执行写操作。尽管如此，系统仍然有可能在读线程之前调度写线程，此时缓冲区是空的。因此，每次读线程获得一些内容，就要把读取的字节数加给 `dwBytesRed` 变量，只有写线程的字节数小于读线程的字节数，才可以执行写操作。否则，本轮循环将跳过写操作，并释放互斥量供其他线程使用。

## 更多讨论

生产者-消费者问题也称为有界缓冲问题。两个进程共享一个固定大小的公共缓冲区。生产者把信息放入缓冲区，消费者把信息从缓冲区中取出来。该问题也可扩展为 m 个生产者和 n 个消费者的问题。不过，这里我们简化了问题，只考虑一个生产者和一个消费者。当生产者往缓冲区放入新项目时缓冲区满了，就会产生问题。解决的方案是，让生产者睡眠，并在消费者已经移除一个或多个项时唤醒生产者。同理，如果消费者要从缓冲区取项目时缓冲区为空，也会产生问题。解决的方案是，让消费者睡眠，等生产者把项目放入缓冲区后再唤醒它。这种方法看起来很简单，但是会导致竞态条件。读者可以用学过的知识，尝试解决类似的情况。

## 2.10 在内核实现线程

整个内核就是一个进程,许多系统(内核)线程在其上下文中运行。内核有一个线程表,跟踪该系统中所有的线程。

内核维护这个传统的进程表以跟踪进程。那些可以阻塞线程的函数调用可作为系统调用执行,这比执行系统过程的代价更高。当线程被阻塞时,内核必须运行其他线程。当线程被毁坏时,则被标记为不可运行。但是,它的内核数据结构不会受到影响。然后在创建新的线程时,旧的线程将被再次激活,回收资源以备后用。当然,也可以回收用户级线程,但如果线程管理开销非常小,就没必要这样做。

### 准备就绪

下面的示例要求安装 WinDDK(*Driver Development Kit*,驱动程序开发工具包),详情请参阅附录。成功安装 WinDDK 后,运行 Visual Studio。

### 操作步骤

1. 创建一个新的 Win32 应用程序项目,并命名为 `KernelThread`。

2. 打开【解决方案资源管理器】,并添加一个新的头文件,命名为 `ThreadApp.h`。打开 `ThreadApp.h`,并输入下面的代码:

   ```
   #include <windows.h>
   #include <tchar.h>

   #define DRIVER_NAME TEXT( "TestDriver.sys" )
   #define DRIVER_SERVICE_NAME TEXT( "TestDriver" )
   #define Message(n) MessageBox(0, TEXT(n), \
   TEXT("Test Driver Info"), 0)

   BOOL StartDriver(LPTSTR szCurrentDriver);
   BOOL StopDriver(void);
   ```

3. 现在,打开【解决方案资源管理器】,并添加一个新的源文件,命名为 `ThreadApp.cpp`。打开 `ThreadApp.cpp`,并输入下面的内容:

   ```
   #include "ThreadApp.h"

   int APIENTRY WinMain(HINSTANCE hInstance, HINSTANCE hPrevInstance,
       LPSTR szCommandLine, int iCmdShow)
   {
       StartDriver(DRIVER_NAME);
       ShellAbout(0, DRIVER_SERVICE_NAME, TEXT(""), NULL);
       StopDriver();
       return 0;
   }
   ```

```c
BOOL StartDriver(LPTSTR szCurrentDriver)
{
    HANDLE hFile = 0;
    DWORD dwReturn = 0;
    SC_HANDLE hSCManager = { 0 };
    SC_HANDLE hService = { 0 };
    SERVICE_STATUS ServiceStatus = { 0 };
    TCHAR szDriverPath[MAX_PATH] = { 0 };
    GetSystemDirectory(szDriverPath, MAX_PATH);
    TCHAR szDriver[MAX_PATH + 1];
#ifdef _UNICODE
    wsprintf(szDriver, L"\\drivers\\%ws", DRIVER_NAME);
#else
    sprintf(szDriver, "\\drivers\\%s", DRIVER_NAME);
#endif
    _tcscat_s(szDriverPath, (_tcslen(szDriver) + 1) * sizeof(TCHAR),
        szDriver);
    BOOL bSuccess = CopyFile(szCurrentDriver, szDriverPath, FALSE);
    if (bSuccess == FALSE)
    {
        Message("copy driver failed");
        return bSuccess;
    }
    hSCManager = OpenSCManager(NULL, NULL,
        SC_MANAGER_CREATE_SERVICE);
    if (hSCManager == 0)
    {
        Message("open sc manager failed!");
        return FALSE;
    }
    hService = CreateService(hSCManager, DRIVER_SERVICE_NAME,
        DRIVER_SERVICE_NAME, SERVICE_START | DELETE | SERVICE_STOP,
        SERVICE_KERNEL_DRIVER, SERVICE_DEMAND_START, SERVICE_ERROR_IGNORE,
        szDriverPath, NULL, NULL, NULL, NULL, NULL);
    if (hService == 0)
    {
        hService = OpenService(hSCManager, DRIVER_SERVICE_NAME,
            SERVICE_START | DELETE | SERVICE_STOP);
        Message("create service failed!");
    }
    if (hService == 0)
    {
        Message("open service failed!");
        return FALSE;
    }
    BOOL startSuccess = StartService(hService, 0, NULL);
    if (startSuccess == FALSE)
    {
        Message("start service failed!");
        return startSuccess;
    }
    CloseHandle(hFile);
    return TRUE;
}

BOOL StopDriver(void)
{
    SC_HANDLE hSCManager = { 0 };
    SC_HANDLE hService = { 0 };
```

```
        SERVICE_STATUS ServiceStatus = { 0 };
        TCHAR szDriverPath[MAX_PATH] = { 0 };
        GetSystemDirectory(szDriverPath, MAX_PATH);
        TCHAR szDriver[MAX_PATH + 1];
#ifdef _UNICODE
        wsprintf(szDriver, L"\\drivers\\%ws", DRIVER_NAME);
#else
        sprintf(szDriver, "\\drivers\\%s", DRIVER_NAME);
#endif
        _tcscat_s(szDriverPath, (_tcslen(szDriver) + 1) * sizeof(TCHAR),
            szDriver);
        hSCManager = OpenSCManager(NULL, NULL,
            SC_MANAGER_CREATE_SERVICE);
        if (hSCManager == 0)
        {
            return FALSE;
        }
        hService = OpenService(hSCManager, DRIVER_SERVICE_NAME,
            SERVICE_START | DELETE | SERVICE_STOP);
        if (hService)
        {
            ControlService(hService, SERVICE_CONTROL_STOP,
                &ServiceStatus);
            DeleteService(hService);
            CloseServiceHandle(hService);
            BOOL ifSuccess = DeleteFile(szDriverPath);
            return TRUE;
        }
        return FALSE;
}
```

4. 现在,打开【解决方案资源管理器】,创建一个新的空 Win32 控制台项目,并命名为 DriverApp。

5. 添加一个新的头文件,命名为 DriverApp.h,并输入以下代码:

```
#include <ntddk.h>

DRIVER_INITIALIZE DriverEntry;
DRIVER_UNLOAD OnUnload;
```

6. 打开【解决方案资源管理器】,在 DriverApp 项目下,添加一个新的源文件,命名为 DriverApp.cpp。打开 DriverApp.cpp,并输入以下代码:

```
#include "DriverApp.h"

VOID ThreadStart(PVOID lpStartContext)
{
    PKEVENT pEvent = (PKEVENT)lpStartContext;
    DbgPrint("Hello! I am kernel thread. My ID is %u. Regards..",
        (ULONG)PsGetCurrentThreadId());
    KeSetEvent(pEvent, 0, 0);
    PsTerminateSystemThread(STATUS_SUCCESS);
}

NTSTATUS DriverEntry(PDRIVER_OBJECT theDriverObject, PUNICODE_STRING
   theRegistryPath)
{
```

```
            HANDLE hThread = NULL;
            NTSTATUS ntStatus = 0;
            OBJECT_ATTRIBUTES ThreadAttributes;
            KEVENT kEvent = { 0 };
            PETHREAD pThread = 0;
            theDriverObject->DriverUnload = OnUnload;
            DbgPrint("Entering KERNEL mode..");
            InitializeObjectAttributes(&ThreadAttributes, NULL, OBJ_KERNEL_HANDLE,
                NULL, NULL);
            __try
            {
                KeInitializeEvent(&kEvent, SynchronizationEvent, 0);
                ntStatus = PsCreateSystemThread(&hThread, GENERIC_ALL,
                    &ThreadAttributes, NULL, NULL, (PKSTART_ROUTINE)&ThreadStart,
                    &kEvent);
                if (NT_SUCCESS(ntStatus))
                {
                    KeWaitForSingleObject(&kEvent, Executive, KernelMode, FALSE,
                        NULL);
                    ZwClose(hThread);
                }
                else
                {
                    DbgPrint("Could not create system thread!");
                }
            }
            __except (EXCEPTION_EXECUTE_HANDLER)
            {
                DbgPrint("Error while creating system thread!");
            }
            return STATUS_SUCCESS;
        }

        VOID OnUnload(PDRIVER_OBJECT DriverObject)
        {
            DbgPrint("Leaving KERNEL mode..");
        }
```

### 示例分析

首先，我们创建了一个 Win32 应用程序（仅作为演示用），没有 UI，也没有消息循环。我们只想把驱动程序加载至内核中。然后，程序将显示 ShellAbout 对话框，这仅仅是为了让用户有时间阅读 DbgView 输出（欲详细了解 DbgView，请参阅附录）。在用户关闭 ShellAbout 对话框后，程序将卸载驱动程序，应用程序结束。

我们创建的 Win32 应用程序只能加载和卸载驱动程序，所以不做进一步解释了。现在，来看 DriverApp 项目。为编译驱动程序设置好 Visual Studio 后（请查阅附录了解详细的 Visual Studio 的编译设置），我们声明了下面两个主例程，每个驱动程序都必须在 DriverApp.h 头文件中：

```
DRIVER_INITIALIZE DriverEntry;
DRIVER_UNLOAD OnUnload;
```

这两个例程是驱动程序入口点和驱动程序的卸载例程。我们将使用驱动程序入口点初始化一个线程对象，并启动内核线程。新创建的线程将只写入一条显示它唯一标识符的消息，然后立刻返回。要说明的是，深入探讨和开发内核超出了本书讨论的范围，我们在这里浅尝辄止。因为驱动程序被编译为/TC2，我们必须确保在执行第一条命令之前已经声明了所有变量。如下代码所示：

```
HANDLE hThread = NULL;
NTSTATUS ntStatus = 0;
OBJECT_ATTRIBUTES ThreadAttributes;
KEVENT kEvent = { 0 };
PETHREAD pThread = 0;
```

然后，还必须设置卸载例程：

```
theDriverObject->DriverUnload = OnUnload;
```

另外，在创建内核线程之前，要用 `InitializeObjectAttributes` 例程初始化 `ThreadAttribute` 对象：

```
InitializeObjectAttributes(&ThreadAttributes, NULL, OBJ_KERNEL_HANDLE, NULL, NULL);
```

内核开发必须执行得非常谨慎，哪怕是一丁点儿错误都会导致**蓝屏死机**（*BSOD*）或机器崩溃。为此，我们使用 `__try - __except` 块，它与我们熟悉的 `try - catch` 块稍有不同。

在内核中创建句柄和在用户空间中创建句柄不同。`KeWaitForSingleObject` 例程无法使用 `PsCreateSystemThread` 返回的句柄。我们要在 `KeWaitForSingleObject` 返回时在线程中初始化一个触发的事件（事件将在第 3 章中详细介绍）。`PsCreateSystemThread` 例程必须与 `ZwClose` 例程成对调用。调用 `ZwClose` 关闭内核句柄和防止内存泄漏。

最后，我们要实现 `PKSTART_ROUTINE` 或线程的开始地址，线程的指令从这里开始执行。下面是一个示例：

```
VOID (__stdcall* KSTART_ROUTINE)( PVOID StartContext );
```

我们已经通过 `PsCreateSystemThread` 的最后一个参数传递了一个指向 `KEVENT` 的指针。现在，使用 `DbgPrint` 把相应的线程 ID 写入消息供用户阅读。然后，设置一个事件，以便相应的 `KeWaitForSingleObject` 调用可以返回并安全地退出驱动程序。确保 `PsTerminateSystemThread` 没有返回。内核将在卸载驱动程序时清理线程对象。

## 更多讨论

虽然内核线程能解决一些问题，但也不是万能的。例如，当多线程进程创建其他多线程进程时会发生什么情况？应该创建与旧进程的线程一样多的新进程，还是创建只有一个线程的新进程？在多数情况下，这取决于你下一步打算用进程做什么。如果要启动一个新程序，也许应该创建只有一个线程的进程。但如果是继续执行，也许应该创建具有同样数量线程的进程才对。

# 第 3 章 管理进程

本章介绍以下内容：
- ▶ 进程和线程
- ▶ 协作式任务处理和抢占式任务处理
- ▶ 解释 Windows 线程对象
- ▶ 基本线程管理
- ▶ 实现未同步线程
- ▶ 使用同步线程
- ▶ Win32 同步对象和同步技术

## 3.1 介绍

要更好地管理和使用线程，我们要理解 3 个主要的线程操作：线程创建、线程终止和加入父线程。如前所述，每个进程必须至少有 1 个线程，即**主线程**。因此，这里涉及线程的情况，也适用于进程。

每个线程都能创建其他线程。创建线程用 Win32 API 的 CreateThread，下面是 CreateThread API 的原型。

```
HANDLE WINAPI CreateThread(
    LPSECURITY_ATTRIBUTES lpThreadAttributes,
    SIZE_T dwStackSize,
    LPTHREAD_START_ROUTINE lpStartAddress,
    LPVOID lpParameter,
    DWORD dwCreationFlags,
    LPDWORD lpThreadId
);
```

通过创建单独的线程，我们把要在单独实体上并发或不并发执行的任务和操作进行划分。也就是说，既可以不阻塞调用方（创建方）异步运行单独的线程，也可以在调用方被阻塞时（等待线程完成相应的操作[通常称之为等待返回]）同步运行单独的线程。每次要执行并发任务时，都要设法把完成任务所需的操作进行划分才能并发地执行这些操作。我们可以通过创建单独的线程来完成。

和其他任务一样，线程不会永无止尽地执行下去。当创建一个线程时，必须提供它的 StartAddress。StartAddress 是一个指向用户定义例程的指针，线程从这个例程开始执行。当例程返回时，该线程终止。下面是一个 StartAddress 的示例：

```
DWORD WINAPI StartAddress(
    LPVOID lpParameter
);
```

用 Win32 API 的 `TerminateThread` 可以强制终止线程。但是，以这种方式终止线程并不正确，如无必要请勿使用。因为这样强行终止线程，操作系统不会妥善释放线程使用的资源。

线程被终止后，系统会释放其占用的资源，并设置线程的退出码。线程被终止时将触发线程对象，如果进程中只有一个线程，那么该进程也将被终止。

强行终止线程可能会陷入麻烦，例如未释放内存栈、未清理资源或未通知相关联的 DLL。而且，被终止的线程所持有的句柄也未关闭。

许多情况下，线程都需要彼此合作。例如，如果线程 A 需要从线程 B 获得输出数据（用户输入或类似输入），我们常说线程 A 要等待线程 B。当线程 A 要等待线程 B 时，我们说线程 B 要加入线程 A。当一个执行流在另一个执行流执行完毕之后才能继续执行时，就属于这种情况。

## 3.2 进程和线程

确定何时使用进程、何时使用线程非常重要。首先要意识到，进程必须要从磁盘加载一个文件。必须理解进程是完整的实体。另外，创建进程要使用许多系统资源。别忘了，进程还需其他东西，如守护程序（daemon）、监听器（服务器）等。因此，应该在真正需要的时候才使用进程。

而创建线程比创建进程快得多，开支也小得多。而且，创建线程不用从磁盘加载，只要提供 `StartAddress` 指针就行了，这是一个用户定义的例程（函数）。因此，要进行小型的并发操作时，绝大多数情况下都应该选择线程。

我们在前面章节的示例中使用过进程和线程，下面的示例将演示进程和线程在创建和使用方面的巨大区别。

### 准备就绪

确定安装并运行了 Visual Studio。

### 操作步骤

现在，根据以下步骤创建程序，稍后再详细解释。

1. 创建一个新的 C++控制台应用程序，并命名为 `tmpThread`。

2. 打开 `tmpThread.cpp`。

3. 添加下面的代码：

```cpp
#include "stdafx.h"
#include <windows.h>
#include <iostream>

using namespace std;

DWORD WINAPI StartAddress(LPVOID lpParameter)
{
    cout << "Hello. I am a very simple thread."
        << endl
        << "I am used to demonstrate thread creation."
        << endl;
    return 0;
}

int _tmain(int argc, _TCHAR* argv[])
{
    // 创建线程。注意，我们只需在相同的代码段中定义 StartAddress 例程即可，
    // 因为主函数负责开始线程和执行并发操作。
    HANDLE hThread = CreateThread(NULL, 0, StartAddress, NULL, 0, NULL);

    // 创建进程。注意 CreateProcess 的第一个参数，它指向要从磁盘加载的文件。
    STARTUPINFO startupInfo = { 0 };
    PROCESS_INFORMATION processInformation = { 0 };

#ifdef _DEBUG
    BOOL bSuccess = CreateProcess(
        TEXT("..//Debug//tmpProcess.exe"), NULL, NULL,
        NULL, FALSE, 0, NULL, NULL, &startupInfo,
        &processInformation);
#else
    BOOL bSuccess = CreateProcess(
        TEXT("..//Release//tmpProcess.exe"), NULL, NULL,
        NULL, FALSE, 0, NULL, NULL, &startupInfo,
        &processInformation);
#endif

    WaitForSingleObject(hThread, INFINITE);
    CloseHandle(hThread);

    return system("pause");
}
```

4. 创建一个新的 C++ 控制台应用程序，命名为 `tmpProcess`。

5. 打开【解决方案资源管理器】。在 `tmpThread` 项目上单击右键，选择【项目依赖项】，在下拉菜单中选择 `tmpThread`，然后选择下面的 `tmpProcess` 项目。

6. 打开 `tmpProcess.cpp`。

7. 添加下面的代码：

```cpp
#include "stdafx.h"
```

```
#include <iostream>
int _tmain(int argc, _TCHAR* argv[])
{
    std::cout << "Hello. I am a very simple process."
        << std::endl
        << "I am used to demonstrate process creation."
        << std::endl;

    return 0;
}
```

### 示例分析

上面的示例非常简单，目的是为了演示创建进程和创建线程的区别。在同一个代码段中实现一个例程（可以在单独的代码段或者甚至在 DLL 中实现）就能创建并使用线程，而管理并行线程是一项非常复杂的工作。在本例中，我们把进程的创建和使用定义为并发。通过等待进程的主线程可以实现同步，只不过是在线程级执行，而不是在进程级执行。

我们换一种方式来解释进程和线程的最大区别。在使用进程时，进程是一个完整的实体，有地址空间、对象上下文、文件句柄等。然而，进程很贪婪，它想占用所有的资源。进程的地址是私有的，其他进程无法使用某一进程的地址（指针），这使得进程之间的通信较难实现。

在图 3.1 中，3 个进程加载至物理内存中，准备执行或正在 CPU 上执行。

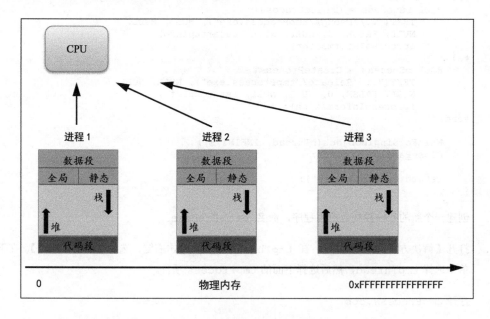

图 3.1　3 个进程加载至物理内存中

如果要频繁地处理进程的通信和同步，最好选择线程。与进程相比，线程更加友好。如果进行合理地管理和同步，线程也可以提供并行，其执行速度比进程快。线程的优点是，所有的进程资源都属于共同的地址空间。句柄、指针和对象都可以使用这些资源，这大幅减少了从 RAM 转移至 CPU 的系统开销。

如图 3.2 所示，一个有 3 个线程的进程被加载至内存中。它的线程准备在 CPU 上执行。

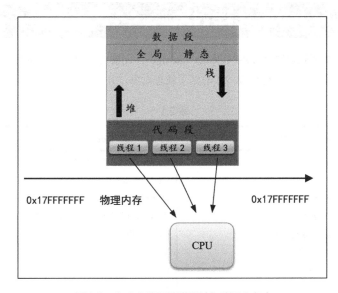

图 3.2　有 3 个线程的进程被加载至内存中

### 更多讨论

每次程序要执行异步操作，都应该考虑创建线程。例如，有多个窗口的程序比单个窗口的程序创建的线程多。内置自动语法检查文字处理器（例如，微软的 Word）的情况也是如此。一个（或多个）线程负责响应键盘的动作，一个线程更新屏幕上的文本，一个线程定期更新工作文件的副本，等等。

## 3.3　协作式和抢占式多任务处理

学习线程的行为有助于理解当前操作系统的实现。接下来，我们简要介绍一下多任务处理。

在 Windows 2000 之前的 Windows 操作系统用的是协作式多任务处理原则，后来的 Windows 操作系统实现了抢占式多任务处理原则。

在协作式多任务处理中，操作系统依赖资源的平均共享。也就是说，应用程序把控制权返回操作系统，以便让其他进程（应用程序）也能获得处理器时间。

在抢占式多任务处理中，操作系统按时暂停线程的执行，让其他线程也能获得相等的处理器时间。

显然，抢占式多任务处理在执行任务时更简单、更安全。实际上，控制权返回变得那么自然而然，既不用苦等进程"吃饱喝足后的恩赐"，也不用程序员为此大伤脑筋。

## 3.4 解释 Windows 线程对象

在操作系统级别，线程是**对象管理器**（Object Manager）创建的一个对象。和所有系统对象类似，线程有自己的特征（数据）和方法（函数）。表 2.1 总结了线程对象，并列出了它的特征和方法。

表 2.1

| Object attributes（对象特征） | Object methods（对象方法） |
| --- | --- |
| Client ID（客户 ID） | Create thread（创建线程） |
| Context（上下文） | Open thread（打开线程） |
| Dynamic priority（动态优先级） | Query thread information（询问线程信息） |
| Base priority（基本优先级） | Set thread information（设置线程信息） |
| Process affinity（进程关联性） | Current thread（当前线程） |
| Execution time（执行时间） | Terminate thread（终止线程） |
| Alert status（警告状态） | Get context（获得上下文） |
| Suspension count（挂起计数） | Set context（设置上下文） |
| Impersonation symbol（模拟象征） | Suspend（挂起） |
| Termination port（终止端口） | Resume（恢复） |
| Exit status（退出状态） | Alert（警告） |
| | Test alert（测试警告） |
| | Register termination port（注册终止端口） |

大部分线程的方法都有许多 Win32 API 函数。例如，调用 `SuspenThread` 时，Win32 子系统将调用相应的 `Suspend` 方法。也就是说，Win32 API 为 Win32 应用程序调用了 `Suspend` 方法。

Windows 总是要保护一些内部结构，如直接操控的窗口或画刷。在用户级别执行的线程不能直接检查或修改系统对象的内部，只能通过调用合适的 Win32 API 方法才行。这些方法可以访问系统对象。Windows 提供了用于识别系统对象的识别码，程序员把识别码传递给所需的函数即可。

线程一般都有识别码，如信号量、事件、文件和所有的对象的识别码。只有对象管理器可以改变对象的状态。`CreateThread` Win32 API 例程创建的线程会为新对象返回一个识别码。有了识别码，我们就可以执行下面的任务：

- 提高或降低线程的优先级；

- 暂停或继续运行线程；
- 终止线程；
- 找出线程的退出码。

## 3.5 基本线程管理

抽象线程对于解决特定的问题很有用。在本节，我们将实现一个线程类，让读者对线程实现的抽象有一个大概的了解。为了方便执行同步，还将实现一个辅助类 CLock。

### 准备就绪

确定安装并运行了 **Visual Studio**。

### 操作步骤

现在，根据以下步骤创建程序，稍后再详细解释。

1. 创建一个新的空 C++ 控制台应用程序，并命名为 CThread。
2. 添加一个新的头文件 CThread.h。
3. 在 CThread.h 中添加下面的代码：

```cpp
#ifndef _CTHREAD_
#define _CTHREAD_

#include <windows.h>

#define STATE_RUNNING     0x0001
#define STATE_READY       0x0002
#define STATE_BLOCKED     0x0004
#define STATE_ALIVE       0x0008
#define STATE_ASYNC       0x0010
#define STATE_SYNC        0x0020
#define STATE_CONTINUOUS  0x0040

class CLock
{
public:
    CLock();
    ~CLock();
private:
    HANDLE hMutex;
};

class CThread
{
public:
    CThread() : hThread(0), dwThreadId(0), dwState(0),
        lpUserData(0), lpParameter(0){ }
```

```cpp
    HANDLE Create(LPVOID lpParameter,
        DWORD dwInitialState = STATE_ASYNC, DWORD dwCreationFlag = 0);
    void Join(DWORD dwWaitInterval = INFINITE);
    DWORD Suspend();
    DWORD Resume();
    void SetUserData(void* lpUserData);
    void* GetUserData() const;
    DWORD GetId() const;
    HANDLE GetHandle() const;
    DWORD GetAsyncState() const;
    DWORD GetState() const;
    void SetState(DWORD dwNewState);
    BOOL Alert();

protected:
    virtual void Run(LPVOID lpParameter = 0) = 0;
    LPVOID lpParameter;

private:
    static DWORD WINAPI StartAddress(LPVOID lpParameter);
    HANDLE hThread;
    DWORD dwThreadId;
    DWORD dwState;
    void* lpUserData;
    HANDLE hEvent;
};

inline DWORD CThread::GetId() const
{
    return dwThreadId;
}

inline HANDLE CThread::GetHandle() const
{
    return hThread;
}

#endif
```

4. 添加一个新的源文件 CThread。打开 CThread.cpp 文件,添加下面的代码:

```cpp
#include "CThread.h"

CLock::CLock()
{
    hMutex = CreateMutex(NULL, FALSE, TEXT("_tmp_mutex_lock_"));
    WaitForSingleObject(hMutex, INFINITE);
}

CLock::~CLock()
{
    ReleaseMutex(hMutex);
    CloseHandle(hMutex);
}

HANDLE CThread::Create(LPVOID lpParameter, DWORD dwInitialState,
    DWORD dwCreationFlag)
{
    dwState |= dwInitialState;
```

```cpp
        this->lpParameter = lpParameter;
        if (dwState & STATE_ALIVE)
        {
            return hThread;
        }
        hThread = CreateThread(NULL, 0, StartAddress, this,
            dwCreationFlag, &dwThreadId);
        dwState |= STATE_ALIVE;
        if (dwState & STATE_CONTINUOUS)
        {
            hEvent = CreateEvent(NULL, TRUE, FALSE, TEXT("__tmp_event__"));
        }

        return hThread;
    }

    void CThread::Join(DWORD dwWaitInterval)
    {
        if (dwState & STATE_BLOCKED)
        {
            return;
        }
        if (dwState & STATE_READY)
        {
            return;
        }
        dwState |= STATE_READY;
        WaitForSingleObject(hThread, dwWaitInterval);
        dwState ^= STATE_READY;
    }

    DWORD CThread::Suspend()
    {
        if (dwState & STATE_BLOCKED)
        {
            return DWORD(-1);
        }
        if (dwState & STATE_READY)
        {
            return DWORD(-1);
        }

        DWORD dwSuspendCount = SuspendThread(hThread);
        dwState |= STATE_BLOCKED;

        return dwSuspendCount;
    }

    DWORD CThread::Resume()
    {
        if (dwState & STATE_RUNNING)
        {
            return DWORD(-1);
        }
        DWORD dwSuspendCount = ResumeThread(hThread);
        dwState ^= STATE_BLOCKED;

        return dwSuspendCount;
    }
```

```cpp
void CThread::SetUserData(void* lpUserData)
{
    this->lpUserData = lpUserData;
}

void* CThread::GetUserData() const
{
    return lpUserData;
}

DWORD CThread::GetAsyncState() const
{
    if (dwState & STATE_ASYNC)
    {
        return STATE_ASYNC;
    }
    return STATE_SYNC;
}

DWORD CThread::GetState() const
{
    return dwState;
}

void CThread::SetState(DWORD dwNewState)
{
    dwState = 0;
    dwState |= dwNewState;
}

BOOL CThread::Alert()
{
    return SetEvent(hEvent);
}

DWORD WINAPI CThread::StartAddress(LPVOID lpParameter)
{
    CThread* cThread = (CThread*)lpParameter;
    if (cThread->GetAsyncState() == STATE_SYNC)
    {
        if (cThread->dwState & STATE_CONTINUOUS)
        {
            DWORD dwWaitStatus = 0;
            while (TRUE)
            {
                cThread->Run();

                dwWaitStatus = WaitForSingleObject(cThread->hEvent, 10);

                if (dwWaitStatus == WAIT_OBJECT_0)
                {
                    break;
                }
            }

            return 0;
        }

        cThread->Run();
```

```
            return 0;
        }
        if (cThread->dwState & STATE_CONTINUOUS)
        {
            DWORD dwWaitStatus = 0;
            while (TRUE)
            {
                CLock lock;
                {
                    cThread->Run();
                }

                dwWaitStatus = WaitForSingleObject(cThread->hEvent, 10);

                if (dwWaitStatus == WAIT_OBJECT_0)
                {
                    break;
                }
            }
            return 0;
        }
        CLock lock;
        {
            cThread->Run();
        }
        return 0;
    }
```

5. 添加一个新的源文件 main.cpp，并输入下面的代码：

```
#include "CThread.h"

class Thread : public CThread
{
protected:
    virtual void Run(LPVOID lpParameter = 0);
};

LRESULT CALLBACK WindowProcedure(HWND hWnd, UINT uMsg, WPARAM
    wParam, LPARAM lParam);

void Thread::Run(LPVOID lpParameter)
{
    WNDCLASSEX wndEx = { 0 };

    wndEx.cbClsExtra = 0;
    wndEx.cbSize = sizeof(WNDCLASSEX);
    wndEx.cbWndExtra = 0;
    wndEx.hbrBackground = (HBRUSH)COLOR_BACKGROUND;
    wndEx.hCursor = LoadCursor(NULL, IDC_ARROW);
    wndEx.hIcon = LoadIcon(NULL, IDI_APPLICATION);
    wndEx.hIconSm = LoadIcon(NULL, IDI_APPLICATION);
    wndEx.hInstance = GetModuleHandle(NULL);
    wndEx.lpfnWndProc = WindowProcedure;
    wndEx.lpszClassName = (LPCTSTR) this->lpParameter;
    wndEx.lpszMenuName = NULL;
    wndEx.style = CS_HREDRAW | CS_VREDRAW | CS_DBLCLKS;
```

```cpp
            if (!RegisterClassEx(&wndEx))
            {
                return;
            }

            HWND hWnd = CreateWindow(wndEx.lpszClassName,
                TEXT("Basic Thread Management"), WS_OVERLAPPEDWINDOW, 200,
                200, 800, 600, HWND_DESKTOP, NULL, wndEx.hInstance, NULL);

            UpdateWindow(hWnd);
            ShowWindow(hWnd, SW_SHOW);

            MSG msg = { 0 };
            while (GetMessage(&msg, 0, 0, 0))
            {
                TranslateMessage(&msg);
                DispatchMessage(&msg);
            }

            UnregisterClass(wndEx.lpszClassName, wndEx.hInstance);
}

int main()
{
    Thread thread;
    thread.Create(TEXT("WND_CLASS_1"));

    Thread pthread;
    pthread.Create(TEXT("WND_CLASS_2"));

    thread.Join();
    pthread.Join();

    return 0;
}

LRESULT CALLBACK WindowProcedure(HWND hWnd, UINT uMsg,
    WPARAM wParam, LPARAM lParam)
{
    switch (uMsg)
    {
        case WM_DESTROY:
        {
            PostQuitMessage(0);
            break;
        }
        case WM_CLOSE:
        {
            DestroyWindow(hWnd);
            break;
        }
        default:
        {
            return DefWindowProc(hWnd, uMsg, wParam, lParam);
        }
    }
    return 0;
}
```

### 示例分析

我们先仔细分析 `CThread.h` 头文件。`CThread` 类是带有纯虚成员函数（`Run`）的抽象类。创建抽象类并不是要直接使用它，而是为了派生用。首先，声明了 `CLock` 类，在需要执行同步时要用到该类，它的实现在 `CThread.cpp` 文件中。在 `CLock` 构造函数中，我们要创建一个互斥量，或打开一个现有互斥量。由于调用 `CreateMutex` API 时传入的第 2 个参数是 `FALSE`，所以一旦线程创建了互斥量，该线程就不再拥有其所有权。线程要调用 `WaitForSingleObject` 函数才能拥有互斥量的所有权。如果另有线程也需要互斥量，那么在互斥量被释放之前，这个需要互斥量的线程将被阻塞。待操作完成，`CLock` 被释放，其析构函数被调用。在 `CLock` 的析构函数中，将释放互斥量和关闭互斥量的句柄。

对于 `CThread` 类，我们为线程状态和执行方法定义了 7 个宏。

- ▶ `STATE_RUNNING`：设置运行标志，说明线程正在运行。
- ▶ `STATE_READY`：设置就绪标志，说明线程正在加入调用方。
- ▶ `STATE_BLOCKED`：设置阻塞标志，说明线程被挂起。
- ▶ `STATE_ALIVE`：设置激活标志，说明线程被创建。
- ▶ `STATE_ASYNC`：设置异步标志，说明线程将并发执行，即在线程返回之前，调用方不会阻塞线程。
- ▶ `STATE_SYNC`：设置同步标志，说明线程将不会并发执行，即调用方在线程返回之前将一直阻塞。
- ▶ `STATE_CONTINUOUS`：设置连续标志，说明线程将连续工作永不停止，或者至少在进程退出前一直运行。

`CThread` 类构造函数的实现非常简单，只要把该类特征的值设置为 0 就行了。`Create` 方法创建实际的线程。必须提供第 1 个参数 `lpParameter`，稍后该参数将被传给 `StartAddress` 方法，线程从这里开始执行。第 2 个参数 `dwAsyncState`，代表线程执行方法。该参数的默认值为 `STATE_ASYNC`，但是用户在需要时可将其改为 `STATE_SYNC` 或 `STATE_CONTINOUS`。如果设置为 `STATE_SYNC`，就使用 `CLock` 类对象。第 3 个参数 `dwCreationFlag` 默认设置线程状态为 0，即在创建该线程后立即使用它；如果设置为 `STATE_BLOCKED` 标志，说明将创建一个处于挂起状态的线程，这相当于 `CREATE_SUSPENDED`；如果设置为 `STATE_CONTINUOUS` 标志，说明将创建事件，如果该事件已存在，则打开事件对象。该标志意味着必须以常规方式运行连续任务，只要所属的进程存在就不要终止它。

`Join` 方法设置 `STATE_READY` 标志，并等待线程对象成为已触发状态，或者在 `dwWaitInterval` 时间间隔内一直等待。默认情况下，`dwWaitInterval` 参数设置为 `INFINITE`。

`Suspend` 和 `Resume` 方法分别挂起和恢复线程。`STATE_BLOCKED` 标志在 `Suspend` 中设置，在 `Resume` 中原原。线程类设计了 `lpUserData` 特征，为的是将来方便设置用户指定的数据。`SetUserData` 和 `GetUserDate` 方法分别设置和获取用户的数据值。`GetAsyncState` 方法返回对象的异步状态。

`GetState` 和 `SetState` 方法分别获取和设置 `dwState` 值或对象状态，用于处理额外的对象状态。

`Alert` 方法把事件的状态改为 `signaled`，用于处理使用了 `STATE_CONTINUOUS` 的情况。如果线程本应运行连续任务，就必须有某些机制以常规的方式停止线程。所以，当我们有一个或多个线程连续运行时，只需从一个线程实例中调用 `Alert`，就能向所有连续运行的线程发警报，停止继续运行并返回。

由于类成员函数（方法）的第 1 个参数都是隐式参数 `this`（即该参数不显示），所以必须把 `StartAddress` 方法声明为 `static`。尽管如此，我们可以把对象的地址（`this` 指针）传递给 `StartAddress` 方法，使用纯虚函数 `Run` 使其行为像一个非静态方法。

`GetId` 和 `GetHandle` 方法分别返回一个唯一的线程标识符和一个线程句柄。这两个方法都实现为内联例程。

### 更多讨论

前面的示例中，我们遵循面向对象设计方案，在实现线程的过程中体现了简洁的编程风格和代码复用。下一节涉及的 `CThread` 用法会稍难。

## 3.6 实现异步的线程

在许多情况下，并行任务要各自独立完成。有多用户界面对话框的应用程序就是这样。例如，在主应用程序窗口中有一个显示客户的列表视图，在其他应用程序窗口（或应用程序选项卡）中有一个显示产品的列表视图。当我们开始运行这样的应用程序时，它会依次先加载客户列表视图，再加载产品列表视图。我们可以启用两个线程从数据库中加载对象（一个线程加载客户，一个线程加载产品）。

现在，来看一个简单的示例。我们要初始化主应用程序窗口，然后创建 4 个子窗口，并为每个窗口创建一个单独的线程，不限制系统资源（在本例中，表示每个窗口都有填色的画刷）。稍后，在下一个示例中将限制只有一个画刷可用，用于解释同步动作。

### 准备就绪

确定安装并运行了 **Visual Studio**。

### 操作步骤

现在，根据下面的步骤创建程序，稍后再详细解释。

1. 创建一个新的空 Win32 项目，并命名为 `AsyncThreads`。

2. 添加一个名为 `AsyncThreads` 的新源文件。打开 `AsyncThreads.cpp`。

3. 输入下面的代码：

```
#include "CThread.h"
#include <windowsx.h>
#include <tchar.h>
#include <time.h>

#define THREADS_NUMBER    4
#define WINDOWS_NUMBER    10
#define SLEEP_TIME        500

LRESULT CALLBACK WindowProcedure(HWND hWnd, UINT uMsg, WPARAM
    wParam, LPARAM lParam);

class Thread : public CThread
{
protected:
    virtual void Run(LPVOID lpParameter = 0);
};

void Thread::Run(LPVOID lpParameter)
{
    // 随机种子生成器
    srand((unsigned)time(NULL));
    HWND hWnd = (HWND)GetUserData();
    RECT rect;

    // 获得窗口的尺寸
    BOOL bError = GetClientRect(hWnd, &rect);
    if (!bError)
    {
        return;
    }

    int iClientX = rect.right - rect.left;
    int iClientY = rect.bottom - rect.top;

    // 如果窗口没有尺寸，不绘图
    if ((!iClientX) || (!iClientY))
    {
        return;
    }

    // 获得设备上下文以绘图
    HDC hDC = GetDC(hWnd);

    if (hDC)
    { // 绘制10个随机图形
        for (int iCount = 0; iCount < WINDOWS_NUMBER; iCount++)
        {
            // 设置坐标
            int iStartX = (int)(rand() % iClientX);
            int iStopX = (int)(rand() % iClientX);
            int iStartY = (int)(rand() % iClientY);
            int iStopY = (int)(rand() % iClientY);
            // 设置颜色
            int iRed = rand() & 255;
            int iGreen = rand() & 255;
```

```
                    int iBlue = rand() & 255;

                    // 创建画刷
                    HANDLE hBrush = CreateSolidBrush(GetNearestColor(hDC,
                        RGB(iRed, iGreen, iBlue)));
                    HANDLE hbrOld = SelectBrush(hDC, hBrush);

                    Rectangle(hDC, min(iStartX, iStopX), max(iStartX, iStopX),
                        min(iStartY, iStopY), max(iStartY, iStopY));

                    // 删除画刷
                    DeleteBrush(SelectBrush(hDC, hbrOld));
                }

                // 释放设备上下文
                ReleaseDC(hWnd, hDC);
        }

        Sleep(SLEEP_TIME);

        return;
    }

    int WINAPI _tWinMain(HINSTANCE hThis, HINSTANCE hPrev, LPTSTR
        szCommandLine, int iCmdShow)
    {
        WNDCLASSEX wndEx = { 0 };

        wndEx.cbClsExtra = 0;
        wndEx.cbSize = sizeof(WNDCLASSEX);
        wndEx.cbWndExtra = 0;
        wndEx.hbrBackground = (HBRUSH)COLOR_BACKGROUND;
        wndEx.hCursor = LoadCursor(NULL, IDC_ARROW);
        wndEx.hIcon = LoadIcon(NULL, IDI_APPLICATION);
        wndEx.hIconSm = LoadIcon(NULL, IDI_APPLICATION);
        wndEx.hInstance = hThis;
        wndEx.lpfnWndProc = WindowProcedure;
        wndEx.lpszClassName = _T("async_thread");
        wndEx.lpszMenuName = NULL;
        wndEx.style = CS_HREDRAW | CS_VREDRAW | CS_DBLCLKS;

        if (!RegisterClassEx(&wndEx))
        {
            return 1;
        }

        HWND hWnd = CreateWindow(wndEx.lpszClassName,
            TEXT("Basic Thread Management"), WS_OVERLAPPEDWINDOW, 200,
            200, 840, 440, HWND_DESKTOP, NULL, wndEx.hInstance, NULL);

        HWND hRects[THREADS_NUMBER];

        hRects[0] = CreateWindow(_T("STATIC"), _T(""), WS_BORDER |
            WS_CHILD | WS_VISIBLE | WS_CLIPCHILDREN, 20, 20, 180,
            350, hWnd, NULL, hThis, NULL);
        hRects[1] = CreateWindow(_T("STATIC"), _T(""), WS_BORDER |
            WS_CHILD | WS_VISIBLE | WS_CLIPCHILDREN, 220, 20, 180,
            350, hWnd, NULL, hThis, NULL);
        hRects[2] = CreateWindow(_T("STATIC"), _T(""), WS_BORDER |
```

```
            WS_CHILD | WS_VISIBLE | WS_CLIPCHILDREN, 420, 20, 180,
            350, hWnd, NULL, hThis, NULL);
        hRects[3] = CreateWindow(_T("STATIC"), _T(""), WS_BORDER |
            WS_CHILD | WS_VISIBLE | WS_CLIPCHILDREN, 620, 20, 180,
            350, hWnd, NULL, hThis, NULL);

        UpdateWindow(hWnd);
        ShowWindow(hWnd, SW_SHOW);

        Thread threads[THREADS_NUMBER];
        for (int iIndex = 0; iIndex < THREADS_NUMBER; iIndex++)
        {
            threads[iIndex].Create(NULL, STATE_ASYNC | STATE_CONTINUOUS);
            threads[iIndex].SetUserData(hRects[iIndex]);
        }

        SetWindowLongPtr(hWnd, GWLP_USERDATA, (LONG_PTR)threads);

        MSG msg = { 0 };
        while (GetMessage(&msg, 0, 0, 0))
        {
            TranslateMessage(&msg);
            DispatchMessage(&msg);
        }

        UnregisterClass(wndEx.lpszClassName, wndEx.hInstance);
        return 0;
    }

    LRESULT CALLBACK WindowProcedure(HWND hWnd, UINT uMsg, WPARAM
        wParam, LPARAM lParam)
    {
        switch (uMsg)
        {
        case WM_DESTROY:
        {
            PostQuitMessage(0);
            break;
        }
        case WM_CLOSE:
        {
            Thread* pThread = (Thread*)GetWindowLongPtr(hWnd,
                GWLP_USERDATA);
            pThread->Alert();
            for (int iIndex = 0; iIndex < THREADS_NUMBER; iIndex++)
            {
                pThread[iIndex].Join();
            }
            DestroyWindow(hWnd);
            break;
        }
        default:
        {
            return DefWindowProc(hWnd, uMsg, wParam, lParam);
        }
        }
        return 0;
    }
```

4. 把 CThread.h 和 CThread.cpp 复制到储存该项目的目录中。

5. 打开【解决方案资源管理器】,在【头文件】上单击鼠标右键,选择【添加】,然后选择【现有项】,添加 CThread.h 文件。

6. 打开【解决方案资源管理器】,在【源文件】上单击鼠标右键,选择【添加】,然后选择【现有项】,添加 CThread.cpp 文件。

## 示例分析

该示例使用了上一个示例已实现的 CThread 类。如前所述,该类用于派生子类。因此,我们创建了 Thread 类作为 CThread 类的派生类。另外,程序中的 Run 例程是虚函数,所以必须要定义(实现)它。

先解释 Run 例程。首先,计算矩形窗口的大小。要获得坐标才能确定绘制的窗口,为此,我们使用了 GetClientRect API。然后,创建合适的画刷为窗口填色。现在,还需要设备上下文才能进行绘制,我们使用了 GetDC API。除了调用 GetNearestColor 获取已知颜色外,还要调用 CreateSolidBrush。使用完这些资源后,还要调用合适的方法释放它们(例如,分别调用 DeleteBrush 和 ReleaseDC)。

接下来,我们解释主例程。首先,创建一个 WNDCLASSEX 对象,创建和初始化应用程序窗口要用到它,调用 RegisterClassEx 注册应用程序窗口也要用到它。通过 CreateWindows API 创建一个窗口后,还要创建 4 个子窗口,线程将在其中执行绘图动作。调用 UpdateWindows 和 ShowWindow 后,将创建 4 个线程对象。接下来,调用线程对象的方法 Create,把用运算符 OR(|)组合的 STATE_ASYNC 和 STATE_CONTINUOUS 标志作为第 2 个参数。然后,设置用户数据指针指向子窗口句柄,稍后用于识别要绘制的窗口部分。

然后,调用 SetWindowLongPtr。该例程用 GWLP_USERDATA 标志把用户数据与第 1 个参数标识的窗口(本例指主应用程序窗口)相关联。第 3 个参数指的是需要与窗口相关联的特定用户数据。应用程序进入消息循环后,我们将传入 Thread 数组的地址,供 WindowProcedure 稍后使用。

在 WindowProcedure 中只处理两种情况:用户关闭应用程序时和窗口要被销毁时。case WM_CLOSE 中就是我们需要等待线程安全加入的地方。首先,通过调用 GetWindowLongPtr 获得之前已储存的线程数组地址。如前所述,只需要一个线程调用 Alert 方法即可,所有监听事件触发的线程都将获得该地址。然后,为每个线程调用 Join 并退出程序。

## 更多讨论

读者也许注意到了,我们只是根据实际情况在之前定义好的 CThread 类中补充了一些新的设计,非常简单。再次提醒读者,如果需要复用代码,面向对象方案很好用。

## 3.7 实现同步的线程

与 3.6 节的示例不同,许多情况中的并行任务也相互依赖。一个应用程序的用户界面中要显示客户列表和已选客户支付记录,就属于这种情况。例如,在一个应用程序窗口中,上部分显示客户列表视图,下部分显示已选定客户的历史支付记录视图。当加载最初的一组客户时,要显示第 1 位客户的支付记录(加载时已选定的项)。我们启动一个线程加载客户列表,同时还要启动另一个线程加载指定客户的支付记录。在加载完第 1 位客户之前,负责加载支付历史的线程必须等待客户 ID。

现在,我们回到上一个示例。其他情况不变,但是这次限制了系统资源。我们要初始化主应用程序窗口,然后需要创建 4 个子窗口,并为每个窗口创建一个单独的线程。这次,只能用一个画刷给窗口填色。

### 准备就绪

确定安装并运行了 Visual Studio。

### 操作步骤

现在,根据下面的步骤创建程序,稍后再详细解释。

1. 创建一个新的空 C++ Win32 项目,并命名为 `SyncThreads`。

2. 添加一个新的源文件,命名为 `SyncThreads`。打开 `SyncThreads.cpp`。

3. 输入下面的代码:

```cpp
#include "CThread.h"
#include <windowsx.h>
#include <tchar.h>
#include <time.h>

#define THREADS_NUMBER 4
#define WINDOWS_NUMBER 10
#define SLEEP_TIME 500

LRESULT CALLBACK WindowProcedure(HWND hWnd, UINT uMsg, WPARAM
    wParam, LPARAM lParam);

class Thread : public CThread
{
protected:
    virtual void Run(LPVOID lpParameter = 0);
};

void Thread::Run(LPVOID lpParameter)
{
    // 随机种子生成器
    srand((unsigned)time(NULL));
```

```cpp
        HWND hWnd = (HWND)GetUserData();
        RECT rect;

        // 获得窗口的尺寸
        BOOL bError = GetClientRect(hWnd, &rect);
        if (!bError)
        {
            return;
        }
        int iClientX = rect.right - rect.left;
        int iClientY = rect.bottom - rect.top;

        // 如果窗口没有尺寸,不绘图
        if ((!iClientX) || (!iClientY))
        {
            return;
        }

        // 获得设备上下文
        HDC hDC = GetDC(hWnd);

        if (hDC)
        { // 绘制10个随机图形
            for (int iCount = 0; iCount < WINDOWS_NUMBER; iCount++)
            {
                // 设置坐标
                int iStartX = (int)(rand() % iClientX);
                int iStopX = (int)(rand() % iClientX);
                int iStartY = (int)(rand() % iClientY);
                int iStopY = (int)(rand() % iClientY);
                // 设置颜色
                int iRed = rand() & 255;
                int iGreen = rand() & 255;
                int iBlue = rand() & 255;

                // 创建一个实心画刷
                HANDLE hBrush = CreateSolidBrush(GetNearestColor(hDC,
                    RGB(iRed, iGreen, iBlue)));
                HANDLE hbrOld = SelectBrush(hDC, hBrush);

                Rectangle(hDC, min(iStartX, iStopX), max(iStartX, iStopX),
                    min(iStartY, iStopY), max(iStartY, iStopY));

                // 删除画刷
                DeleteBrush(SelectBrush(hDC, hbrOld));
            }

            // 释放设备上下文
            ReleaseDC(hWnd, hDC);
        }

        Sleep(SLEEP_TIME);

        return;
    }

    int WINAPI _tWinMain(HINSTANCE hThis, HINSTANCE hPrev, LPTSTR
        szCommandLine, int iCmdShow)
```

```cpp
{
    WNDCLASSEX wndEx = { 0 };

    wndEx.cbClsExtra = 0;
    wndEx.cbSize = sizeof(WNDCLASSEX);
    wndEx.cbWndExtra = 0;
    wndEx.hbrBackground = (HBRUSH)COLOR_BACKGROUND;
    wndEx.hCursor = LoadCursor(NULL, IDC_ARROW);
    wndEx.hIcon = LoadIcon(NULL, IDI_APPLICATION);
    wndEx.hIconSm = LoadIcon(NULL, IDI_APPLICATION);
    wndEx.hInstance = hThis;
    wndEx.lpfnWndProc = WindowProcedure;
    wndEx.lpszClassName = _T("async_thread");
    wndEx.lpszMenuName = NULL;
    wndEx.style = CS_HREDRAW | CS_VREDRAW | CS_DBLCLKS;

    if (!RegisterClassEx(&wndEx))
    {
        return 1;
    }

    HWND hWnd = CreateWindow(wndEx.lpszClassName,
        TEXT("Basic Thread Management"), WS_OVERLAPPEDWINDOW, 200,
        200, 840, 440, HWND_DESKTOP, NULL, wndEx.hInstance, NULL);

    HWND hRects[THREADS_NUMBER];
    hRects[0] = CreateWindow(_T("STATIC"), _T(""), WS_BORDER |
        WS_CHILD | WS_VISIBLE | WS_CLIPCHILDREN, 20, 20, 180, 350,
        hWnd, NULL, hThis, NULL);
    hRects[1] = CreateWindow(_T("STATIC"), _T(""), WS_BORDER |
        WS_CHILD | WS_VISIBLE | WS_CLIPCHILDREN, 220, 20, 180, 350,
        hWnd, NULL, hThis, NULL);
    hRects[2] = CreateWindow(_T("STATIC"), _T(""), WS_BORDER |
        WS_CHILD | WS_VISIBLE | WS_CLIPCHILDREN, 420, 20, 180, 350,
        hWnd, NULL, hThis, NULL);
    hRects[3] = CreateWindow(_T("STATIC"), _T(""), WS_BORDER |
        WS_CHILD | WS_VISIBLE | WS_CLIPCHILDREN, 620, 20, 180, 350,
        hWnd, NULL, hThis, NULL);

    UpdateWindow(hWnd);
    ShowWindow(hWnd, SW_SHOW);

    Thread threads[THREADS_NUMBER];
    for (int iIndex = 0; iIndex < THREADS_NUMBER; iIndex++)
    {
        threads[iIndex].Create(NULL, STATE_SYNC | STATE_CONTINUOUS);
        threads[iIndex].SetUserData(hRects[iIndex]);
    }

    SetWindowLongPtr(hWnd, GWLP_USERDATA, (LONG_PTR)threads);

    MSG msg = { 0 };
    while (GetMessage(&msg, 0, 0, 0))
    {
        TranslateMessage(&msg);
        DispatchMessage(&msg);
    }

    UnregisterClass(wndEx.lpszClassName, wndEx.hInstance);
```

```
            return 0;
        }
        LRESULT CALLBACK WindowProcedure(HWND hWnd, UINT uMsg, WPARAM
            wParam, LPARAM lParam)
        {
            switch (uMsg)
            {
                case WM_DESTROY:
                {
                    PostQuitMessage(0);
                    break;
                }
                case WM_CLOSE:
                {
                    Thread* pThread = (Thread*)GetWindowLongPtr(hWnd,
                        GWLP_USERDATA);
                    pThread->Alert();
                    for (int iIndex = 0; iIndex < THREADS_NUMBER; iIndex++)
                    {
                        pThread[iIndex].Join();
                    }
                    DestroyWindow(hWnd);
                    break;
                }
                default:
                {
                    return DefWindowProc(hWnd, uMsg, wParam, lParam);
                }
            }
            return 0;
        }
```

4. 把 CTread.h 和 CThread.cpp 复制到该项目的目录中。

5. 打开【解决方案资源管理器】,在【头文件】上单击鼠标右键,选择【添加】,然后选择【现有项】,添加 CThread.h 文件。

6. 打开【解决方案资源管理器】,在【源文件】上单击鼠标右键,选择【添加】,然后选择【现有项】,添加 CThread.cpp 文件。

## 示例分析

该示例的代码和上一个示例的代码几乎完全相同,只有一个字母的区别。即在创建线程时,我们用 STATE_ASYNC 标志而不是用 STATE_SYNC 标志初始化线程。

这两个示例要做的事情都设计在 CThread 类中了。

## 更多讨论

我们在这里要强调的是,强大的面向对象技术以及合理设计解决方案和复用代码会让程序更加简洁。在实际动手开始编程之前，要养成多思考的好习惯。在敲代码之前，多花些时间设计解决方案比直接写更好。再加上面向对象技术，编程就轻松多了。

## 3.8 Win32 同步对象和技术

线程必须被调度才能执行。为了在执行过程中互不干扰，必须进行同步。

假设一个线程先创建一把画刷，再创建多个线程共享该画刷，并用它绘图。在其他线程绘制完成之前不能销毁第 1 个线程。又如，一个线程接受用户的输入并将其写入文件中，此时另一个线程读取该文件的数据并处理文本。如果写入数据的线程正在进行写入操作，读取数据的线程就不能读取。上面讲到的这两个例子都要求要协调多个线程。

一个解决方案是，创建一个全局的 Boolean 变量 bDone，一个线程用它来触发另一个线程。负责写入的线程可将 bDone 设置为 TRUE，而负责读取的线程在发现指示符 bDone 改变之前执行一个循环。这种机制没问题，但是执行循环的线程耗费了大量的 CPU 时间。其实，Windows 操作系统支持许多同步对象，如：

- **互斥量**对象（相互排斥），类似一个窄门，同一时间只获得一个线程。
- **信号量**对象，类似一个关卡，同一时间允许一定数量的线程通过。
- **事件**对象，播报任何线程都能收听到的公共信号。
- **临界区**对象，类似互斥量，但是只能在一个进程中使用。

这些都是由**对象管理器**创建的系统对象。虽然各同步对象协调不同的交互，但是它们的工作方式相同。要执行协同操作的线程等待某个同步对象的回复，并在接收到回复时恢复操作。为避免浪费 CPU 时间，分发器（*dispatcher*）会把处于等待状态的线程从队列中移除；当获得信号时，分发器再让线程继续执行。

甚至不相关的进程也能共享互斥量、信号量和事件。进程可以像线程那样协调共享对象的活动。有三种分配机制。一种机制是通过继承，当一个进程创建另一个进程时，标记识别码为可继承的，这样新进程将获得父进程识别码的副本，而且只有这些识别码会被传递（继承）。

另外两种机制都要调用一个函数为现有对象创建另一个 ID 码。具体调用哪一个函数取决于我们拥有什么信息。如果有目标进程和源进程的识别码，就使用 DuplicateHandle API；如果知道同步对象名，就能使用一个 Open 函数。两个程序可以事先协商好，何时使用它们共享的同步对象名，或者一个程序通过共享内存或管道引用另一个程序的对象名。

在互斥量、信号量和事件所属的所有进程（即，拥有这些同步对象的进程）执行完毕之前，或者 CloseHandle API 关闭所有同步对象的识别码之前，这些同步对象都驻留在内存中。

## 3.8.1 同步对象：互斥量

如何接收信号以及何时收到信号取决于同步对象。例如，互斥量的一个重要特性是，只允许一个线程拥有它。也就是说，一次只有一个线程获得互斥量的所有权。如果多个线程要处理一个文件，就要创建一个互斥量保护该文件。然后，任何线程在开始操作文件之前，都必须先询问互斥量的所有权。如果没有线程持有互斥量，该线程就拥有互斥量的所有权。如果有线程先获取了互斥量，这样的询问就失败。而且，这个询问互斥量的线程被阻塞，在等待获得互斥量所有权时将被挂起。当线程写入完毕时将释放互斥量，正在等待的线程一接收到互斥量便立即恢复，继续在文件中执行所需的操作。

互斥量并不保护任何激活的线程，它的工作方式是：线程必须先持有互斥量才能使用共享对象。但是这阻止不了线程尝试立即注册。互斥量是一个信号，类似前面例子中的 Boolean 变量 bDone，可用于保护全局变量、硬件端口、管道的识别码，或窗口的客户区。为了避免相互冲突，当多个线程要共享系统资源时都应该考虑使用互斥量来同步。

下面是一些互斥量的使用例程。创建或打开一个现有的互斥量对象，可以使用 CreateMutex：

```
HANDLE WINAPI CreateMutex(
    LPSECURITY_ATTRIBUTES    lpMutexAttributes,
    BOOL                     bInitialOwner,
    LPCTSTR                  szName
);
```

第 1 个参数是安全特征，可以指定也可以空出不填（如果不填，该参数将被设为默认值）。第 2 个参数表示所有权，如果把 bInitialOwner 设置为 TRUE，调用方将立即获得互斥量的所有权。第 3 个参数指定互斥量的名字，必须提供互斥量名。

用 OpenMutex 也可以打开一个现有的互斥量：

```
HANDLE WINAPI OpenMutex(
    DWORD dwDesiredAccess,
    BOOL bInheritHandle,
    LPCTSTR szName
);
```

如果互斥量不存在，该例程失败并返回 ERROR_FILE_NOT_FOUND。线程操作完毕后，必须释放互斥量以便让其他线程获取。可以使用 ReleaseMutex API 释放互斥量：

```
BOOL WINAPI ReleaseMutex(
    HANDLE hMutex
);
```

该例程只有一个参数，即之前 CreateMutex 或 OpenMutex 返回的互斥量句柄。

## 3.8.2 同步对象：信号量

如前所述，信号量与互斥量类似，唯一的区别是多个对象可持有信号量的所有权。假设有一个复杂的数学运算要用到整个逻辑 CPU 核。如果每个核只运行一个线程，计算结果没问题；但是，如果每个核运行多个线程，计算结果就有问题了。另外，还假设计算过程中所需的线程数量比同时工作的逻辑核的数量多。

最好选择信号量对象来处理这种情况，我们可以把信号量对象的最大值设置为机器的逻辑核数量。当线程数量不超过核的数量时，它们能同时工作，优化计算过程。当线程数量超过逻辑核的数量时，一些线程将被挂起，等待其他线程执行完毕。

每次只有一个线程可以获得互斥量，与此不同的是，只要未超过可持有信号量所有权的最大数量，信号量仍然处于触发状态。如果一个线程要等待信号量，那么在其他线程释放信号量之前它将被挂起。

下面是一些信号量的使用例程。创建或打开一个现有的信号量对象，可以使用 CreateSemaphore：

```
HANDLE WINAPI CreateSemaphore(
    LPSECURITY_ATTRIBUTES    lpSemaphoreAttributes,
    LONG                     lInitialCount,
    LONG                     lMaximumCount,
    LPCTSTR                  szName
);
```

第 1 个参数是安全特征，可以指定也可以空出不填（如果不填，该参数将被设为默认值）。第 2 个参数设置初始数目以确定信号量的初始触发状态。如果设置为 0，信号量为未触发状态。这个值必须小于第 3 个参数 lMaxmumCount 的值。lMaxmumCount 表示可同时持有信号量所有权的最多对象（线程）数目。必须提供信号量名（szName）。

使用 OpenSemaphore 也可以打开一个现有的信号量：

```
HANDLE WINAPI OpenSemaphore(
    DWORD dwDesiredAccess,
    BOOL bInheritHandle,
    LPCTSTR szName
);
```

如果信号量不存在，该例程将失败并返回 NULL。线程操作完毕后，必须释放信号量以递减计数器计数，这样其他线程才能获得信号量。用 ReleaseSemaphore API 可以释放信号量：

```
BOOL WINAPI ReleaseSemaphore(
    HANDLE hSemaphore,
    LONG lReleaseCount,
    LPLONG lpPreviousCount
);
```

这个 API 有 3 个参数。第 1 个参数是之前 CreateSemaphore 或 OpenSemaphore 返回的信号量的句柄。第 2 个参数是信号量要递减的对象数量，其值通常是 1，因为线程总是逐个被释放的。有一种情况例外：

线程 A 获得一个信号量，然后要创建也需要信号量的线程 B。线程 A 在线程 B 创建好之前完成了自己的任务，线程 B 不知道它将被强制终止，所以线程 A 要终止子线程且递减信号量为 2。

## 3.8.3 同步对象：事件

程序在发生某些动作时，需要某种机制警告线程，事件就是为此而创建的对象。手动重置事件是最简单的形式，这种形式的事件对象用 SetEvent（信号开）和 ResetEvent（信号关）两个命令来分别设置信号开和信号关作为响应。当信号开启时，所有等待事件的线程都将收到信号；当信号关闭时，所有等待事件的线程都将被阻塞。与互斥量和信号量不同，只有线程显式设置和重置事件，事件才能改变自身的状态。

例如，如果要求某些线程只在程序不绘制窗口时，或者只在用户输入某些信息后才运行，就可以使用手动重置事件。

自动重置事件在重置之前，每次只会释放一个线程。当程序的主线程为其他工作线程准备数据时，自动重置事件很有用。每次准备好一组新数据，主线程就设置事件并释放工作线程。而其他工作线程仍在队列中等待获取新任务。

除了设置和重置事件，还可以脉冲事件。脉冲是在非常短的时间内设置信号，然后关闭信号。当脉冲一个手动重置事件时，我们允许所有正在等待的线程继续执行，然后重置事件。当脉冲一个自动重置事件时，只允许一个正在等待的线程通过，然后重置事件。如果没有线程等待，就不会有线程通过。另一方面，设置一个自动事件将导致事件在线程等待时触发。一旦一个线程通过，事件就被重置。

下面是一些事件的使用例程。创建或打开一个现有的事件对象，可以使用 CreateEvent 或 CreateEventEx：

```
HANDLE WINAPI CreateEvent(
    LPSECURITY_ATTRIBUTES    lpEventAttributes,
    BOOL                     bManualReset,
    BOOL                     bInitialState,
    LPCTSTR                  szName
);

HANDLE WINAPI CreateEventEx(
    LPSECURITY_ATTRIBUTES    lpEventAttributes,
    LPCTSTR                  szName,
    DWORD                    dwFlags,
    DWORD                    dwDesiredAccess
);
```

对于 CreateEvent API，可以指定安全特征，也可以空出不填（如果不填，该参数将被设为默认值）。bManualReset 值用于确定事件对象。如果 bManualReset 设置为 TRUE，则必须把 ResetEvent 中的事件对象设置为未触发状态。bInitialState 值设置事件的初始状态。如果设置为 TRUE，该事件对象则被立即设置为已触发状态。必须提供事件名（szName）。

CreateEventEx 与 CreateEvent 的工作原理几乎相同，只是描述的术语不同。其中描述的安全特征和事件名的参数一样。对于 dwFlag 的名称和值，系统为初始设置和手动重置事件提供了预定义宏：CREATE_EVENT_INITIAL_SET 和 CREATE_EVENT_MANUAL_RESET。这两个 API 唯一真正的区别是 CreateEventEx 的 dwDesiredAccess 参数可以给事件对象设置访问掩码。欲详细了解同步对象安全和访问权限，请查阅 MSDN（http://msdn.microsoft.com/en-us/library/windows/desktop/ms686670%28v=vs.85%29.aspx）。

还可以用 OpenEvent 打开一个现有的事件对象：

```
HANDLE WINAPI OpenEvent(
    DWORD dwDesiredAccess,
    BOOL bInheritHandle,
    LPCTSTR szName
);
```

如果事件不存在，该例程失败并返回 NULL。调用 OpenEvent 可以让多线程持有相同的事件，但前提是调用 CreateEvent 或 CreateEx 创建该事件。

用 SetEvent 可以设置事件对象：

```
BOOL WINAPI SetEvent(
    HANDLE hEvent
);
```

SetEvent 只有一个参数，即之前用 CreateEvent 和 CreateEventEx 创建或由 OpenEvent 打开的事件对象的句柄。手动重置事件在调用 ResetEvent API 显式设置为未触发状态之前一直保持已触发状态。

重置事件对象，可以使用 ResetEvent：

```
BOOL WINAPI ResetEvent(
    HANDLE hEvent
);
```

ResetEvent 也只有一个参数，即之前用 CreateEvent 和 CreateEventEx 创建或由 OpenEvent 打开的事件对象的句柄。手动重置事件在调用 SetEvent API 显式设置为已触发状态之前一直保持未触发状态。

用 PulseEvent 可以脉冲一个事件对象：

```
BOOL WINAPI PulseEvent(
    HANDLE hEvent
);
```

该方法把事件对象设置为已触发状态，然后将其重置为未触发状态，释放特定数量的等待线程。

### 3.8.4 同步对象：临界区

临界区执行的功能与互斥量相同，不同的是临界区不能共享，它只对一个进程可见。临界区和互斥量都只允许一个线程拥有它,但是临界区运行速度更快,开销更小。

虽然操作临界区的例程表面上与使用互斥量的例程不同,但是它们所做的事情几乎相同。

和其他需要等待函数来等待的同步对象不同,临界区对象的工作方式不一样。一开始,临界区对象处于未触发状态;在被释放后,临界区对象才进入触发状态。

要先声明临界区对象才能使用它:

```
CRITICAL_SECTION cs;
```

接着,要初始化它:

```
void WINAPI InitializeCriticalSection(
    LPCRITICAL_SECTION lpCriticalSection
);
```

然后,通过调用 EnterCriticalSection 或 TryEnterCriticalSection API 让一个线程进入临界区:

```
void WINAPI EnterCriticalSection(
    LPCRITICAL_SECTION lpCriticalSection
);

BOOL WINAPI TryEnterCriticalSection(
    LPCRITICAL_SECTION lpCriticalSection
);
```

线程完成任务后,必须调用 LeaveCriticalSection API 离开临界区:

```
void WINAPI LeaveCriticalSection(
    LPCRITICAL_SECTION lpCriticalSection
);
```

接着,还要调用 DeleteCriticalSection API 释放资源:

```
void WINAPI DeleteCriticalSection(
    LPCRITICAL_SECTION lpCriticalSection
);
```

# 第 4 章 消息传递

**本章介绍以下内容：**
- 解释消息传递接口
- 理解消息队列
- 使用线程消息队列
- 通过管道对象进行通信

## 4.1 介绍

要解释**消息传递接口**（*Message Passing Interface*，MPI），首先要了解操作系统。第 1 章中提到过，现代 Windows 操作系统是作为事件驱动系统来设计的。

事件驱动系统将发生某种行为以响应发送给操作系统的信号。例如，当你移动鼠标时，引发了 `cursor X-coordinate` 和 `cursor Y-coordinate change` 事件（我们说引发事件，而不是发生事件）。当你按下键盘上的任意一个键时，就引发了 `key press` 事件等。

在操作系统级别，每几毫秒都会引发大量事件，只是用户觉察不到罢了。如果没有事件，程序设计的难度会很大，特别是很难完成同步任务。

在操作系统级别，消息（*message*）是一种形式的事件。当你按下鼠标左键时，操作系统为窗口区域发生的单击调用窗口过程（事件处理器）。要把定义消息的参数和事件相关的细节（如，单击坐标）传递给窗口过程。

每个带 UI 的应用程序都必须有某种形式的窗口。顶层窗口必须实现顶层窗口处理器或窗口过程，处理应用程序事件（消息），例如，按钮单击、窗口填色等。

为了处理窗口的消息，应用程序必须实现事件处理器。每次引发事件，操作系统都要调用事件处理器。读者应该记得，每个应用程序都必须有自己主线程，因此我们可以用这个主线程来处理其他线程的消息。也就是说，可以从主应用程序窗口使用窗口过程，接收其他线程的消息，彼此通信。

还有其他给线程传递消息的方法，如 `PostThreadMessage` API。但是，寄送（*post*）消息的线程必须创建一个消息队列。我们稍后解释这种方案。尽管如此，如果应用程序有用户界面，使用窗口消息队列更简单。

# 第4章 消息传递

本章将详细介绍如何给不同的任务传递消息,特别是如何在线程间的同步中使用消息传递。

## 4.2 解释消息传递接口

为了详细地解释消息传递,我们先讨论按钮单击。和其他的控件类似,这属于顶层窗口。当引发按钮单击事件时,将调用该窗口过程作为鼠标按下按钮的响应。

下面的示例演示了消息处理的基本用法,窗口过程负责鼠标移动和按钮单击。

### 准备就绪

确定安装并运行了 Visual Studio。

### 操作步骤

现在,根据下面的步骤创建我们的程序,稍后再详细解释。

1. 创建一个新的空 Win32 项目,并命名为 basic_MPI。

2. 打开【解决方案资源管理器】,在【源文件】上单击右键,添加一个新的源文件 basic_MPI。打开 basic_MPI.cpp,并输入下面的代码:

```
#include <windows.h>
#include <windowsx.h>

#include <tchar.h>
#include <commctrl.h>

#pragma comment ( lib, "comctl32.lib" )
#pragma comment ( linker, "\"/manifestdependency:type='win32' \
    name='Microsoft.Windows.Common-Controls' \
    version='6.0.0.0' processorArchitecture='*' \
    publicKeyToken='6595b64144ccf1df' language='*'\"" )

#define BUTTON_MSG 100
#define BUTTON_CLOSE 101
#define LABEL_TEXT 102

LRESULT CALLBACK WindowProcedure(HWND hWnd, UINT uMsg, WPARAM
    wParam, LPARAM lParam);

int WINAPI _tWinMain(HINSTANCE hThis, HINSTANCE hPrev, LPTSTR
    szCommandLine, int iWndShow)
{
    UNREFERENCED_PARAMETER(hPrev);
    UNREFERENCED_PARAMETER(szCommandLine);

    TCHAR* szWindowClass = _T("__basic_MPI_wnd_class__");
```

```
WNDCLASSEX wndEx = { 0 };
wndEx.cbSize = sizeof(WNDCLASSEX);
wndEx.style = CS_HREDRAW | CS_VREDRAW;
wndEx.lpfnWndProc = WindowProcedure;
wndEx.cbClsExtra = 0;
wndEx.cbWndExtra = 0;
wndEx.hInstance = hThis;
wndEx.hIcon = LoadIcon(wndEx.hInstance, MAKEINTRESOURCE(IDI_APPLICATION));
wndEx.hCursor = LoadCursor(NULL, IDC_ARROW);
wndEx.hbrBackground = (HBRUSH)(COLOR_WINDOW + 1);
wndEx.lpszMenuName = NULL;
wndEx.lpszClassName = szWindowClass;
wndEx.hIconSm = LoadIcon(wndEx.hInstance, MAKEINTRESOURCE(IDI_APPLICATION));

if (!RegisterClassEx(&wndEx))
{
    return 1;
}

InitCommonControls();

HWND hWnd = CreateWindow(szWindowClass, _T("Basic Message
    Passing Interface"), WS_OVERLAPPED | WS_CAPTION | WS_SYSMENU |
    WS_MINIMIZEBOX, 200, 200, 440, 340, NULL, NULL, wndEx.hInstance,
    NULL);

if (!hWnd)
{
    return NULL;
}

HFONT hFont = CreateFont(14, 0, 0, 0, FW_NORMAL, FALSE,
    FALSE, FALSE, BALTIC_CHARSET, OUT_DEFAULT_PRECIS,
    CLIP_DEFAULT_PRECIS, DEFAULT_QUALITY, DEFAULT_PITCH
    | FF_MODERN, _T("Microsoft Sans Serif"));

HWND hButtonMsg = CreateWindow(_T("BUTTON"), _T("Show msg"),
    WS_CHILD | WS_VISIBLE | BS_PUSHBUTTON | WS_TABSTOP, 190,
    260, 100, 25, hWnd, (HMENU)BUTTON_MSG, wndEx.hInstance, NULL);

HWND hButtonClose = CreateWindow(_T("BUTTON"), _T("Close me"),
    WS_CHILD | WS_VISIBLE | BS_PUSHBUTTON | WS_TABSTOP, 310, 260,
    100, 25, hWnd, (HMENU)BUTTON_CLOSE, wndEx.hInstance, NULL);

HWND hText = CreateWindow(_T("STATIC"), NULL, WS_CHILD |
    WS_VISIBLE | SS_LEFT | WS_BORDER, 15, 20, 390, 220, hWnd,
    (HMENU)LABEL_TEXT, wndEx.hInstance, NULL);

SendMessage(hButtonMsg, WM_SETFONT, (WPARAM)hFont, TRUE);
SendMessage(hButtonClose, WM_SETFONT, (WPARAM)hFont, TRUE);
SendMessage(hText, WM_SETFONT, (WPARAM)hFont, TRUE);

ShowWindow(hWnd, iWndShow);
UpdateWindow(hWnd);

MSG msg;
while (GetMessage(&msg, 0, 0, 0))
{
    TranslateMessage(&msg);
```

```
            DispatchMessage(&msg);
    }

    UnregisterClass(wndEx.lpszClassName, wndEx.hInstance);

    return (int)msg.wParam;
}

LRESULT CALLBACK WindowProcedure(HWND hWnd, UINT uMsg, WPARAM
    wParam, LPARAM lParam)
{
    switch (uMsg)
    {
        case WM_COMMAND:
        {
            switch (LOWORD(wParam))
            {
                case BUTTON_MSG:
                {
                    MessageBox(hWnd, _T("Hello!"), _T("Basic MPI"), MB_OK |
                        MB_TOPMOST);
                    break;
                }
                case BUTTON_CLOSE:
                {
                    PostMessage(hWnd, WM_CLOSE, 0, 0);
                    break;
                }
            }
            break;
        }

        case WM_MOUSEMOVE:
        {
            int xPos = GET_X_LPARAM(lParam);
            int yPos = GET_Y_LPARAM(lParam);

            TCHAR szBuffer[4096];
            wsprintf(szBuffer,
                _T("\n\n\t%ws\t%d\n\t%ws\t%d"),
                _T("Cursor X position:"), xPos,
                _T("Cursor Y position:"), yPos);

            HWND hText = GetDlgItem(hWnd, LABEL_TEXT);
            SetWindowText(hText, szBuffer);
            break;
        }
        case WM_DESTROY:
        {
            PostQuitMessage(0);
            break;
        }
        default:
        {
            return DefWindowProc(hWnd, uMsg, wParam, lParam);
        }
    }

    return 0;
}
```

## 示例分析

该示例是为了演示基本的消息传递。读者要特别注意在程序前面定义的 `BUTTON_CLOSE` 和 `LABEL_TEXT` 宏。我们在后面的示例中将用作子控件标识符。对于子窗口，hMenu 指定了子窗口标识符。

> 子控件标识符是一个整型值，窗口利用该值把引发的事件告诉父控件。子窗口标识符由应用程序定义，在同一父窗口的所有子窗口中，该标识符必须是独一无二的（摘自 MSDN）。

先来看 WinMain 入口。我们先创建并初始化了 `WNDCLASSEX` 实例。调用 `RegisterClassEx` 成功注册窗口类后，接着调用了 `CreateWindow` 创建了主窗口。

然后，创建了一个文本标签和按钮控件。按钮和标签也是窗口，但是有预定义系统类（`BUTTON`、`STATIC` 等）。欲了解完整的系统类，请参阅 MSDN（http://msdn.microsoft.com/en-us/library/windows/desktop/ms632679(v=vs.85).aspx）。

接着，调用 `ShowWindow`。然后，调用 Win32 API `GetMessage`，应用程序进入 while 中的消息循环。当应用程序关闭顶层窗口时，`GetMessage` 调用返回 0，此时应用程序没有待处理的消息了。

与基于 MS-DOS 的应用程序相比，Win32 应用程序是事件驱动的。它们不会执行显式函数调用来获得输入，而是等待系统把输入传递给它们。操作系统把应用程序指定的所有消息都传递给应用程序窗口。窗口有一个被称为窗口过程的函数，当有窗口的消息时，系统会调用窗口过程。窗口过程负责处理消息和把控制权返回系统。

现在，来看本例的 `WindowProcedure`。我们只处理两条消息：`WM_MOUSEMOVE` 和 `WM_COMMAND`，前者负责鼠标移动，后者负责按钮单击。对于鼠标移动，可以用预定义宏 `GET_X_LPARAM` 获取 x 坐标，用 `GET_Y_LPARAM` 获取 y 坐标。对于按钮单击，我们使用 wParam 参数较低的 16 位来确定哪个子控件引发单击。在确定引发事件的控件后，就可以执行所需的操作。在本例中，**Show msg** 按钮显示一个消息框，**Close me** 按钮关闭应用程序。

## 更多讨论

虽然操作系统提供了专门的同步对象和例程，但是用窗口消息来处理同步任务非常方便。例如，运行一个要异步处理某任务的线程，当工作线程完成任务时，可以通过消息传递通知主线程。在后面的示例中，我们将通过消息传递的方式在线程间进行通信。

## 4.3 理解消息队列

要更好地理解消息传递，先要**了解消息队列**。每个顶层窗口都有属于自己的消息队列，由操作系统创建。在多任务环境中，要同时执行许多应用程序，而且用户的屏幕上也会出现许多窗口。操作系统不仅要处理大量消息，而且还要处理系统本身引发的各种事件以及运行各种应用程序。与此同时，还要处理用户按下键盘上的按键，单击鼠标或滑动触摸屏等。

为了演示消息传递和队列操作，我们来考虑一个例子。假设要绘制一个复杂的四轴图表。每个轴分别代表一种以图样形式表示的数学计算。

在我们创建的应用程序中，一个线程负责计算，一个线程负责根据另一个线程的计算结果绘制图形。假设线程计算时间比绘制图形时间短。

为了给执行计算的线程储存消息，我们还要实现一个消息队列。执行计算的线程（生产者）要比绘制图形的线程（消费者）快，后者处理从消息队列中获得的消息。生产者每次完成任务，就把一个消息放进消息队列中，供消费者稍后处理。

### 准备就绪

确定安装并运行了 Visual Studio。

### 操作步骤

现在，根据下面的步骤创建程序，并解释其结构。

1. 创建一个新的 C++ Win32 空项目，命名为 queue_MPI。把第 1 章中的 CQueue.h 头文件复制到该项目的目录中。

2. 打开【解决方案资源管理器】，右键单击【头文件】，选择【添加】-【现有项】，添加 CQueue.h。

3. 打开【解决方案资源管理器】，右键单击【源文件】，添加一个新的源文件，命名为 queue_MPI。打开 queue_MPI.cpp，输入下面的代码：

   ```
   #include <windows.h>
   #include <windowsx.h>
   #include <tchar.h>
   #include "CQueue.h"
   #include <commctrl.h>

   #pragma comment ( lib, "comctl32.lib" )
   #pragma comment ( linker, "\"/manifestdependency:type='win32'\
   name='Microsoft.Windows.Common-Controls' \
   version='6.0.0.0' processorArchitecture='*' \
   publicKeyToken='6595b64144ccf1df' language='*'\"" )
   ```

```c
#define BUFFER_SIZE         4096

#define LABEL_TEXT          100

#define WM_PLOTDATA     WM_USER         + 1
#define WM_ENDTASK      WM_PLOTDATA     + 1

#define THREAD_NUMBER 4
#define MAX_MSGCOUNT  10

#define CALCULATION_TIME    1000
#define DRAWING_TIME        2300

#define EVENT_NAME          _T( "__t_event__" )

typedef struct tagPLOTDATA
{
    int value;
    DWORD dwThreadId;
    int iMsgID;
} PLOTDATA, *PPLOTDATA;

LRESULT CALLBACK WindowProcedure(HWND hWnd, UINT uMsg, WPARAM wParam,
    LPARAM lParam);
DWORD WINAPI StartAddress(LPVOID lpParameter);
DWORD WINAPI DrawPlot(LPVOID lpParameter);

int iMessageID = 0;
HANDLE gEvent = NULL;
HANDLE hThreads[THREAD_NUMBER];
HANDLE hThread = NULL;

CRITICAL_SECTION cs;
CQueue<PLOTDATA> queue;

int WINAPI _tWinMain(HINSTANCE hThis, HINSTANCE hPrev, LPTSTR szCommandLine,
    int iWndShow)
{
    UNREFERENCED_PARAMETER(hPrev);
    UNREFERENCED_PARAMETER(szCommandLine);

    InitializeCriticalSection(&cs);

    TCHAR* szWindowClass = _T("__basic_MPI_wnd_class__");

    WNDCLASSEX wndEx = { 0 };
    wndEx.cbSize = sizeof(WNDCLASSEX);
    wndEx.style = CS_HREDRAW | CS_VREDRAW;
    wndEx.lpfnWndProc = WindowProcedure;
    wndEx.cbClsExtra = 0;
    wndEx.cbWndExtra = 0;
    wndEx.hInstance = hThis;
    wndEx.hIcon = LoadIcon(wndEx.hInstance, MAKEINTRESOURCE(IDI_APPLICATION));
    wndEx.hCursor = LoadCursor(NULL, IDC_ARROW);
    wndEx.hbrBackground = (HBRUSH)(COLOR_WINDOW + 1);
    wndEx.lpszMenuName = NULL;
    wndEx.lpszClassName = szWindowClass;
    wndEx.hIconSm = LoadIcon(wndEx.hInstance, MAKEINTRESOURCE(IDI_APPLICATION));
```

```cpp
        if (!RegisterClassEx(&wndEx))
        {
            return 1;
        }

        InitCommonControls();

        HWND hWnd = CreateWindow(szWindowClass, _T("Basic Message Passing Interface"),
            WS_OVERLAPPED | WS_CAPTION | WS_SYSMENU | WS_MINIMIZEBOX, 200, 200, 440,
            300, NULL, NULL, wndEx.hInstance, NULL);

        if (!hWnd)
        {
            return NULL;
        }

        HFONT hFont = CreateFont(14, 0, 0, 0, FW_NORMAL, FALSE, FALSE, FALSE,
            BALTIC_CHARSET, OUT_DEFAULT_PRECIS, CLIP_DEFAULT_PRECIS,
            DEFAULT_QUALITY, DEFAULT_PITCH | FF_MODERN, _T("Microsoft Sans Serif"));

        HWND hText = CreateWindow(_T("STATIC"), NULL, WS_CHILD | WS_VISIBLE |
            SS_LEFT | WS_BORDER, 15, 20, 390, 220, hWnd, (HMENU)LABEL_TEXT,
            wndEx.hInstance, NULL);

        SendMessage(hText, WM_SETFONT, (WPARAM)hFont, TRUE);

        ShowWindow(hWnd, iWndShow);

        MSG msg;
        while (GetMessage(&msg, 0, 0, 0))
        {
            TranslateMessage(&msg);
            DispatchMessage(&msg);
        }

        UnregisterClass(wndEx.lpszClassName, wndEx.hInstance);

        WaitForMultipleObjects(THREAD_NUMBER, hThreads, TRUE, INFINITE);

        for (int iIndex = 0; iIndex < THREAD_NUMBER; iIndex++)
        {
            CloseHandle(hThreads[iIndex]);
        }

        WaitForSingleObject(hThread, INFINITE);

        CloseHandle(hThread);
        CloseHandle(gEvent);

        DeleteCriticalSection(&cs);

        return (int)msg.wParam;
}

LRESULT CALLBACK WindowProcedure(HWND hWnd, UINT uMsg, WPARAM wParam, LPARAM lParam)
{
    switch (uMsg)
    {
```

```
            case WM_CREATE:
            {
                gEvent = CreateEvent(NULL, TRUE, FALSE, EVENT_NAME);

                for (int iIndex = 0; iIndex < THREAD_NUMBER; iIndex++)
                {
                    hThreads[iIndex] = CreateThread(NULL, 0,
                        (LPTHREAD_START_ROUTINE)StartAddress, hWnd, 0, NULL);
                }

                hThread = CreateThread(NULL, 0, (LPTHREAD_START_ROUTINE)DrawPlot,
                    hWnd, 0, NULL);
                break;
            }
            case WM_PLOTDATA:
            {
                PPLOTDATA pData = (PPLOTDATA)lParam;

                HWND hLabel = GetDlgItem(hWnd, LABEL_TEXT);

                TCHAR szBuffer[BUFFER_SIZE];
                GetWindowText(hLabel, szBuffer, BUFFER_SIZE);

                wsprintf(szBuffer, _T("%ws\n\n\tMessage has been received. Msg ID:\t%d"),
                    szBuffer, pData->iMsgID);
                SetWindowText(hLabel, szBuffer);

                break;
            }
            case WM_ENDTASK:
            {
                HWND hLabel = GetDlgItem(hWnd, LABEL_TEXT);

                TCHAR szBuffer[BUFFER_SIZE];

                wsprintf(szBuffer,
                    _T("\n\n\tPlot is drawn. You can close the window now."));

                SetWindowText(hLabel, szBuffer);
                break;
            }
            case WM_DESTROY:
            {
                PostQuitMessage(0);
                break;
            }
            default:
            {
                return DefWindowProc(hWnd, uMsg, wParam, lParam);
            }
        }

    return 0;
}

DWORD WINAPI StartAddress(LPVOID lpParameter)
{
    HWND hWnd = (HWND)lpParameter;
```

```cpp
    HANDLE hEvent = OpenEvent(EVENT_ALL_ACCESS, FALSE, EVENT_NAME);

    if (hEvent != NULL)
    {
        SetEvent(hEvent);
    }

    CloseHandle(hEvent);

    int iCount = 0;
    while (iCount++ < MAX_MSGCOUNT)
    {
        // 执行计算
        Sleep(CALCULATION_TIME);

        // 把结果放入 PLOTDATA 结构中
        PPLOTDATA pData = new PLOTDATA();
        pData->value = (rand() % 0xFFFF) - iMessageID;
        pData->dwThreadId = GetCurrentThreadId();
        pData->iMsgID = ++iMessageID;

        EnterCriticalSection(&cs);
        queue.Enqueue(pData);
        LeaveCriticalSection(&cs);

        PostMessage(hWnd, WM_PLOTDATA, 0, (LPARAM)pData);
    }

    return 0L;
}

DWORD WINAPI DrawPlot(LPVOID lpParameter)
{
    HWND hWnd = (HWND)lpParameter;

    HANDLE hEvent = OpenEvent(EVENT_ALL_ACCESS, FALSE, EVENT_NAME);

    WaitForSingleObject(hEvent, INFINITE);
    CloseHandle(hEvent);

    int iCount = 0;
    while (iCount++ < MAX_MSGCOUNT * THREAD_NUMBER)
    {
        EnterCriticalSection(&cs);
        PPLOTDATA pData = queue.Dequeue();
        LeaveCriticalSection(&cs);

        if (!pData)
        {
            break;
        }

        // 执行绘制
        Sleep(DRAWING_TIME);

        HWND hLabel = GetDlgItem(hWnd, LABEL_TEXT);

        TCHAR szBuffer[BUFFER_SIZE];
```

```
            wsprintf(szBuffer,
#ifdef _UNICODE
            _T("\n\n\t%ws\t%u\n\t%ws\t%d\n\t%ws\t%d"),
#else
            _T("\n\n\t%s\t%u\n\t%s\t%d\n\t%s\t%d"),
#endif
            _T("Thread ID:"), (DWORD)pData->dwThreadId,
            _T("Current value:"), (int)pData->value,
            _T("Message ID:"), pData->iMsgID);

            SetWindowText(hLabel, szBuffer);

            delete pData;
        }
        PostMessage(hWnd, WM_ENDTASK, 0, 0);

        return 0L;
    }
```

### 示例分析

首先，我们要定义某些消息值。每个消息都代表一个独一无二的整数，比 Windows 预定义的 WM_USER 值要大。所有比 WM_USER 值（1024）大的消息都是应用程序专用（用户自定义）的。第 1 个 1024 消息给 Windows 专用（预留）。定义两个消息，如下所示。

- ▶ WM_PLOTDATA：负责计算的线程用来通知主线程已完成计算。
- ▶ WM_ENDTASK：用于通知主线程所有的绘制已完成。

在应用程序的入口中，我们创建了一个队列，其中的元素是对 PLOTDATA 类型或某用户自定义结构的引用。在创建窗口后，把下面的队列地址与窗口部分（其中显示用户指定的窗口数据）相关联：

```
SetWindowLongPtr(hWnd, GWLP_USERDATA, (LONG_PTR) &queue);
```

把 GWLP_USERDATA 宏作为 SetWindowLongPtr API 的第 2 个参数，在某种程度上该函数就与队列对象的地址相关联（如果提供窗口句柄 hWnd），这样就可以获得队列的地址。创建必要的控件后，还要创建 4 个线程负责计算和 1 个线程绘制图形。无论何时要用到队列对象的地址，都可以像下面这样用 GetWindowLongPtr 来获取：

```
CQueue<PLOTDATA>* plotPtr = (CQueue<PLOTDATA>*) GetWindowLongPtr( hWnd,GWLP_USERDATA );
```

为了降低绘图的复杂度（这不是我们要讨论的重点），我们将设置工作线程睡眠 1 秒，而绘制线程将睡眠 2.3 秒，目的在于创建一个消费者比生产者慢的环境。所有计算线程要执行固定数量的计算，绘制线程也要执行固定数量的绘制。

每完成一次计算，就创建一个 PLOTDATA 对象并放进队列中。然后，线程调用 PostMessage API 把消息发送（寄送）给第 1 个参数（窗口的句柄）指定的窗口过程。第 2 个参数指定待发送（寄送）的消息，第 3 个和第 4 个参数指定特定信息。如果是 Windows 消息（小于 WM_USER），其含义是已知的；如果是用户自定义的消息，这两个参数就是留给用户使用的。我们使用 `lParam` 参数传递一个新创建的 `PLOTDATA` 对象的地址，以便窗口过程可以显示消息 ID 或用户所需的 `PLOTDATA` 对象中的值。

一方面，生产者创建一条消息后，就把该消息放入消息队列；另一方面，消费者（负责绘制图形的线程）从消息队列中获取消息来执行绘图。如果消费者获得一个指向 `PLOTDATA` 对象的指针（即，队列不为空），则执行绘制。成功绘制后，为了避免内存泄漏，`PLOTDATA` 对象将被销毁。当所有的点绘制完毕、所有的消息处理完毕时，绘图线程将通过 `PostMessage` 通知窗口过程一切执行完毕。

### 更多讨论

虽然我们使用了 `PostMessage` API，但是在描述时仍然用发送消息而不是寄送消息。读者要注意的是，`PostMessage` 和 `SendMessage` 这两个例程都用于发送消息，它们的唯一区别是：`PostMessage` 不用等被调用方处理消息就直接返回，而 `SendMessage` 一直等被调用方处理完消息后才返回。检索消息队列中的消息用 `GetMessage` 或 `PeekMessage` API。

## 4.4 使用线程消息队列

如前所述，不用创建窗口（UI）也可以给线程发送消息。在前面的示例中，我们为属于窗口的线程创建了窗口消息队列。这里，我们将介绍如何不使用窗口消息队列来传递消息。

为了简化示例，我们在每次用户单击按钮时，给线程发送一条消息。

有一个类似的例子，如果要创建一个显示天气预报信息的应用程序，应该有两个线程：一个线程负责物理设备（如，压力表、风速表等）；另一个线程要等待第 1 个线程收集好数据后才能获得数据，然后处理数据并把数据显示给用户。

### 准备就绪

确定安装并运行了 Visual Studio。

### 操作步骤

现在，根据下面的步骤创建程序，并解释其结构。

1. 创建一个新的 C++ Win32 空项目，并命名为 `threadMPI`。把第 1 章中的头文件 `CQueue.h` 复制到该项目的目录中。

2. 打开【解决方案资源管理器】，右键单击【头文件】，选择【添加】-【现有项】，添加 `CQueue.h`。

3. 打开【解决方案资源管理器】，右键单击【源文件】，添加一个新的源文件，命名为 threadMPI。
   打开 threadMPI.cpp，输入下面的代码：

```cpp
#include <windows.h>
#include <windowsx.h>
#include <tchar.h>
#include "CQueue.h"
#include <commctrl.h>

#pragma comment ( lib, "comctl32.lib" )
#pragma comment ( linker, "\"/manifestdependency:type='win32' \
    name='Microsoft.Windows.Common-Controls' \
    version='6.0.0.0' processorArchitecture='*' \
    publicKeyToken='6595b64144ccf1df' language='*'\"" )

#define BUFFER_SIZE     4096

#define BUTTON_CLOSE    100

#define T_MESSAGE       WM_USER     + 1
#define T_ENDTASK       T_MESSAGE   + 1

#define EVENT_NAME      _T( "__t_event__" )

LRESULT CALLBACK WindowProcedure(HWND hWnd, UINT uMsg, WPARAM wParam,
    LPARAM lParam);
DWORD WINAPI StartAddress(LPVOID lpParameter);

int WINAPI _tWinMain(HINSTANCE hThis, HINSTANCE hPrev, LPTSTR szCommandLine,
    int iWndShow)
{
    UNREFERENCED_PARAMETER(hPrev);
    UNREFERENCED_PARAMETER(szCommandLine);

    TCHAR* szWindowClass = _T("__basic_MPI_wnd_class__");

    WNDCLASSEX wndEx = { 0 };
    wndEx.cbSize = sizeof(WNDCLASSEX);
    wndEx.style = CS_HREDRAW | CS_VREDRAW;
    wndEx.lpfnWndProc = WindowProcedure;
    wndEx.cbClsExtra = 0;
    wndEx.cbWndExtra = 0;
    wndEx.hInstance = hThis;
    wndEx.hIcon = LoadIcon(wndEx.hInstance, MAKEINTRESOURCE(IDI_APPLICATION));
    wndEx.hCursor = LoadCursor(NULL, IDC_ARROW);
    wndEx.hbrBackground = (HBRUSH)(COLOR_WINDOW + 1);
    wndEx.lpszMenuName = NULL;
    wndEx.lpszClassName = szWindowClass;
    wndEx.hIconSm = LoadIcon(wndEx.hInstance, MAKEINTRESOURCE(IDI_APPLICATION));

    if (!RegisterClassEx(&wndEx))
    {
        return 1;
    }

    InitCommonControls();

    HWND hWnd = CreateWindow(szWindowClass, _T("Basic Message Passing Interface"),
```

```cpp
            WS_OVERLAPPED | WS_CAPTION | WS_SYSMENU |
            WS_MINIMIZEBOX, 200, 200, 440, 300, NULL, NULL, wndEx.hInstance,
            NULL);

        if (!hWnd)
        {
            return NULL;
        }

        HFONT hFont = CreateFont(14, 0, 0, 0, FW_NORMAL, FALSE,
            FALSE, FALSE, BALTIC_CHARSET, OUT_DEFAULT_PRECIS,
            CLIP_DEFAULT_PRECIS, DEFAULT_QUALITY, DEFAULT_PITCH |
            FF_MODERN, _T("Microsoft Sans Serif"));

        HWND hButton = CreateWindow(_T("BUTTON"), _T("Send message"),
            WS_CHILD | WS_VISIBLE | BS_PUSHBUTTON | WS_TABSTOP, 310, 210, 100,
            25, hWnd, (HMENU)BUTTON_CLOSE, wndEx.hInstance, NULL);

        SendMessage(hButton, WM_SETFONT, (WPARAM)hFont, TRUE);

        ShowWindow(hWnd, iWndShow);

        HANDLE hEvent = CreateEvent(NULL, TRUE, FALSE, EVENT_NAME);

        DWORD dwThreadId = 0;

        HANDLE hThread = CreateThread(NULL, 0, (LPTHREAD_START_ROUTINE)StartAddress,
            hWnd, 0, &dwThreadId);

        SetWindowLongPtr(hWnd, GWLP_USERDATA, (LONG_PTR)&dwThreadId);

        WaitForSingleObject(hEvent, INFINITE);
        CloseHandle(hEvent);

        MSG msg;
        while (GetMessage(&msg, 0, 0, 0))
        {
            TranslateMessage(&msg);
            DispatchMessage(&msg);
        }

        UnregisterClass(wndEx.lpszClassName, wndEx.hInstance);

        PostThreadMessage(dwThreadId, T_ENDTASK, 0, 0);
        WaitForSingleObject(hThread, INFINITE);

        CloseHandle(hThread);
        return (int)msg.wParam;
    }

    LRESULT CALLBACK WindowProcedure(HWND hWnd, UINT uMsg, WPARAM wParam, LPARAM lParam)
    {
        switch (uMsg)
        {
            case WM_COMMAND:
            {
                switch (LOWORD(wParam))
                {
                    case BUTTON_CLOSE:
```

```
                {
                    DWORD* pdwThreadId = (DWORD*)GetWindowLongPtr(hWnd,
                        GWLP_USERDATA);
                    PostThreadMessage(*pdwThreadId, T_MESSAGE, 0, 0);
                    break;
                }
            }
            break;
        }
        case WM_DESTROY:
        {
            PostQuitMessage(0);
            break;
        }
        default:
        {
            return DefWindowProc(hWnd, uMsg, wParam, lParam);
        }
    }
    return 0;
}

DWORD WINAPI StartAddress(LPVOID lpParameter)
{
    MSG msg;
    PeekMessage(&msg, NULL, WM_USER, WM_USER, PM_NOREMOVE);

    HANDLE hEvent = OpenEvent(EVENT_ALL_ACCESS, FALSE, EVENT_NAME);

    if (hEvent != NULL)
    {
        SetEvent(hEvent);
    }
    CloseHandle(hEvent);

    while (GetMessage(&msg, NULL, 0, 0))
    {
        switch (msg.message)
        {
            case T_MESSAGE:
            {
                // 在此添加一些消息处理
                MessageBox(NULL, _T("Hi.\nI've just received a message."),
                    _T("Thread"), MB_OK);
                break;
            }
            case T_ENDTASK:
            {
                // 主线程退出
                // 立刻返回
                return 0L;
            }
        }
    }

    return 0L;
}
```

### 示例分析

与上一个示例一样,每次使用客户消息都必须先定义它。我们定义下面两条消息:

- `T_MESSAGE`:通知线程开始执行它所指定的任务;
- `T_ENDTASK`:通知线程应用程序即将关闭,它要立即返回。

由于没有为新线程创建消息队列,新线程也没有自己的窗口,因此要手动创建。首先,在要发送消息的线程中创建一个事件(生产者)。在新创建的线程创建消息队列之前,不会设置该事件。我们把线程 ID 与窗口关联起来,以便稍后在处理按钮单击时能通过窗口过程使用它。

现在,我们等候线程创建消息队列。以下面的方式调用 `PeekMessage` API,迫使系统创建消息队列:

```
PeekMessage(&msg, NULL, WM_USER, WM_USER, PM_NOREMOVE);
```

创建完消息队列后,可以调用 `SetEvent` API 来设置事件并关闭句柄。然后,线程进入它的消息循环。随后,调用 `GetMessage` 在接收到消息之前阻塞线程。当线程接收到消息时,将检查消息标识符并显示一个消息框。如果是 `T_MESSAGE`,则显示一条消息;否则,在 `T_ENDTASK` 消息的 `case` 中返回。

### 更多讨论

我们在本节中多次提到了生产者和消费者,是为了让读者在普通任务中熟悉生产者/消费者问题。再次强调,一个好的应用程序设计能让整个开发过程事半功倍。设计应用程序时要谨慎、多思考,这是程序开发过程中最重要的一步。

## 4.5 通过管道对象通信

除了消息传递,Windows 还提供其他方式在进程和线程之间通信和共享数据。其中一种方式是**管道**(*pipe*)对象。管道是内存中进程用来通信的对象。创建管道的进程是**管道服务端**(*pipe server*),连接管道的进程是**管道客户端**(*pipe client*)。一个进程把信息写入管道,然后另一个进程从管道中读取信息。

我们要实现一个小型的客户端-服务端示例。一个进程是服务端,一个进程是客户端。服务端将一直运行。我们使用管道对象进行进程通信。每次连接客户端,服务端要同意连接并发送欢迎消息,而客户端要接收服务端的欢迎消息,并且再次发送一条简单的消息向服务端请求。

### 准备就绪

确定安装并运行了 Visual Studio。

### 操作步骤

## 4.5 通过管道对象通信

现在，根据下面的步骤创建程序，稍后再详细解释。

1. 创建一个新的 C++ 控制台应用程序，并命名为 `pipe_server`。

2. 打开【解决方案资源管理器】，右键单击【源文件】，添加一个新的源文件 `pipe_server`。打开 `pipe_server.cpp`。

3. 输入下面的代码：

```cpp
#include <windows.h>
#include <stdio.h>
#include <tchar.h>

#define BUFFER_SIZE 4096

DWORD WINAPI StartAddress(LPVOID lpParameter);

int _tmain(void)
{
    LPTSTR szPipename = _T("\\\\.\\pipe\\basicMPI");

    while (TRUE)
    {
        _tprintf(_T("basicMPI: waiting client connection\n"));

        HANDLE hPipe = CreateNamedPipe(szPipename, PIPE_ACCESS_DUPLEX,
            PIPE_WAIT | PIPE_READMODE_MESSAGE | PIPE_TYPE_MESSAGE,
            PIPE_UNLIMITED_INSTANCES, BUFFER_SIZE, BUFFER_SIZE, 0,
            NULL);

        if (hPipe == INVALID_HANDLE_VALUE)
        {
            _tprintf(_T("CreateNamedPipe failed! Error code: [%u]\n"),
                GetLastError());
            return 2;
        }

        if (ConnectNamedPipe(hPipe, NULL))
        {
            _tprintf(_T("Connection succeeded.\n"));

            HANDLE hThread = CreateThread(NULL, 0,
                (LPTHREAD_START_ROUTINE)StartAddress, hPipe, 0, NULL);

            CloseHandle(hThread);
        }
    }

    return 0;
}

DWORD WINAPI StartAddress(LPVOID lpParameter)
{
    HANDLE hPipe = (HANDLE)lpParameter;

    PTCHAR szRequest = (PTCHAR)HeapAlloc(GetProcessHeap(), 0,
```

```
            BUFFER_SIZE * sizeof(TCHAR));
    PTCHAR szReply = (PTCHAR)HeapAlloc(GetProcessHeap(), 0,
            BUFFER_SIZE * sizeof(TCHAR));
    DWORD dwBytesRead = 0;
    DWORD dwReplyBytes = 0;
    DWORD dwBytesWritten = 0;

    _tprintf(_T("Waiting messages.\n"));

    if (!ReadFile(hPipe, szRequest, BUFFER_SIZE * sizeof(TCHAR),
        &dwBytesRead, NULL))
    {
        _tprintf(_T("ReadFile failed! Error code: [%u]\n"), GetLastError());
        return 2L;
    }

    // 进行一些处理
    _tprintf(_T("Client request:\"%s\"\n"), szRequest);

    _tcscpy_s(szReply, BUFFER_SIZE, _T("default answer from server"));

    dwReplyBytes = (_tcslen(szReply) + 1) * sizeof(TCHAR);

    if (!WriteFile(hPipe, szReply, dwReplyBytes, &dwBytesWritten, NULL))
    {
        _tprintf(_T("WriteFile failed! Error code: [%u]"), GetLastError());
        return 2L;
    }

    FlushFileBuffers(hPipe);
    DisconnectNamedPipe(hPipe);
    CloseHandle(hPipe);

    HeapFree(GetProcessHeap(), 0, szRequest);
    HeapFree(GetProcessHeap(), 0, szReply);

    _tprintf(_T("Success.\n"));

    return 0L;
}
```

4. 打开【解决方案资源管理器】,右键单击【解决方案】,选择【添加】-【新建项目】。创建一个新的空 C++控制台应用程序,命名为 pipe_client。

5. 打开【解决方案资源管理器】,在【源文件】上单击右键,添加一个新的源文件 pipe_client。打开 pipe_client.cpp。

6. 输入下面的代码:

```
#include <windows.h>
#include <stdio.h>
#include <tchar.h>

#define BUFFER_SIZE 4096

int _tmain(void)
```

```
{
    TCHAR szBuffer[BUFFER_SIZE];
    DWORD dwRead = 0;
    DWORD dwWritten = 0;
    DWORD dwMode = PIPE_READMODE_MESSAGE;
    LPTSTR szPipename = _T("\\\\.\\pipe\\basicMPI");
    LPTSTR szMessage = _T("Message from client.");
    DWORD dwToWrite = _tcslen(szMessage) * sizeof(TCHAR);

    HANDLE hPipe = CreateFile(szPipename, GENERIC_READ |
        GENERIC_WRITE, 0, NULL, OPEN_EXISTING, 0, NULL);

    WaitNamedPipe(szPipename, INFINITE);

    SetNamedPipeHandleState(hPipe, &dwMode, NULL, NULL);

    if (!WriteFile(hPipe, szMessage, dwToWrite, &dwWritten, NULL))
    {
        _tprintf(_T("WriteFile failed! Error code: [%u]\n"), GetLastError());
        return -1;
    }

    _tprintf( _T("Message sent to server, receiving reply as follows:\n"));

    while (ReadFile(hPipe, szBuffer, BUFFER_SIZE * sizeof(TCHAR),
        &dwRead, NULL) && GetLastError() != ERROR_MORE_DATA)
    {
        _tprintf(_T("%s"), szBuffer);
    }

    _tprintf(_T("\nSuccess!\nPress ENTER to exit."));
    scanf_s("%c", szBuffer);

    CloseHandle(hPipe);

    return 0;
}
```

### 示例分析

我们先来解释服务端。为了使用管道对象，必须先用 `CreateNamedPipe` API 创建一个。管道必须唯一，必须是 "\\.\pipe\管道名" 的格式。如下所示：

```
HANDLE hPipe = CreateNamedPipe( szPipename, PIPE_ACCESS_DUPLEX,
    PIPE_WAIT | PIPE_READMODE_MESSAGE | PIPE_TYPE_MESSAGE,
    PIPE_UNLIMITED_INSTANCES, BUFFER_SIZE, BUFFER_SIZE, 0, NULL);
```

欲详细了解 `CreateNamedPipe` API，请参阅 MSDN (http://msdn.microsoft.com/en-us/library/windows/desktop/aa365150%28v=vs.85%29.aspx)。

为了处理客户端的请求，服务端必须一直运行着，因此调用 `ConnectNamedPipe` 来执行无限循环，它能让管道服务端等待客户端连接管道。也就是说，如果在 `CreateNamedPipe` 中没有指定 `FILE_FLAG_OVERLAPPED` 就获得了管道对象的句柄，该函数在未连接客户端之前不会返回。客户端连接好后，才继续执行。我们将创建一个处理客户端请求的线程，而主线程则同时准备好下一个请求。

接下来分析请求过程。我们获得管道的句柄后，要为请求和响应分配足够的堆空间。我们使用 ReadFile 从另一个管道的末尾读取，然后进行一些处理。使用 WriteFile 发送响应：

```
WriteFile(hPipe, szReply, dwReplyBytes, &dwBytesWritten, NULL);
```

第 1 个参数是管道句柄；第 2 个参数储存消息缓冲区的地址；第 3 个参数指定待写入的字节数；第 4 个参数是待写入字节缓冲区的地址；最后一个参数是 OVERLAPPED 对象的地址，我们可以用 WriteFile 进行异步操作。写入管道的末尾后，要刷新缓冲区，断开管道连接，并在这里关闭管道句柄，而不是从创建管道的线程关闭。我们这样做是因为主线程为一个客户端打开了管道，如果第 1 个请求尚未处理，又收到了第 2 个请求，主线程就会创建另一个线程处理新的请求。在这种实现中，由处理请求的线程负责打开句柄。

现在，我们分析客户端。有两种方式可以打开管道：CreateFile 或 CallNamedPipe API。客户端在管道可用之前要等待服务端。当服务端调用 ConnectNamedPipe 时继续执行，然后线程被挂起。我们以相同的方式使用 WriteFile 和 ReadFile，就像在服务端处理那样。当所有处理完毕时，必须关闭管道句柄。

### 更多讨论

本章介绍的所有技术都有各自的优缺点。要正确理解这些技术，才能在使用时游刃有余，创造出更多更优的解法。为了让读者熟悉这些技术，下一章将介绍这些技术的实际用法。

# 第 5 章　线程同步和并发操作

本章介绍以下内容：
- 伪并行
- 进程和线程优先级
- Windows 分发器对象和调度
- 使用互斥量
- 使用信号量
- 使用事件
- 使用临界区
- 使用管道

## 5.1　介绍

为了理解不同处理器上的并发执行，先要区分硬件并行和伪并行。现代的处理器都有多个物理内核，这些内核可以并行地处理和计算，而以前的处理器只有一个物理内核，不太可能执行真正的并行。

然后，本章简要介绍 Windows 调度优先级，当我们要执行不同优先级任务时需要了解这些内容。

本章末尾，将使用并比较所有之前介绍的同步技术，总结我们到目前为止学到的内容。

## 5.2　伪并行

前面提到过，现代的处理器有多个物理内核，并行处理是处理器的基本特性。以前的处理器是什么情况？是否能在单核环境中创建并行执行？

举个简单的例子：假设我们要计算一个数学方程式。用一个循环迭代 x 次来计算总和，每次迭代计算一些表达式。

在我们创建的程序中，先用一个线程来执行整个任务，限定计算所需的时间，假设时间为 t。如果把整个任务分成两个线程，第 1 个线程计算前一半迭代的和，第 2 个线程计算后一半迭代的和，执行程序所需的时间是否约为 t/2？

对于单核处理器，不能这样看。因为只能靠操作系统调度执行两个线程，没有硬件支持并发执行。由于只有一个核，CPU 在任意给定时间只能执行一个线程。

提到这些是为了让读者意识到，创建线程有时也会事与愿违。如果我们把任务划分成 8 个线程，然后在单核 CPU 上执行该程序会怎样？显然将拖慢处理速度，可见把任务细分再创建没必要的线程分开执行只是在浪费时间。如果在该程序中添加同步，由于单核 CPU 一次只能运行一个线程，会把大量时间浪费在线程间的等待上。

伪并行是非常重要的操作系统特性。它给用户营造了所有任务（程序）都在并行执行的假象。例如，当你正在使用 Microsoft Word 时，可以播放 Winamp、打印文件，同时还能刻录 DVD。无论是单核还是多核，所有的这些任务都被分成多个单独的线程，由操作系统负责调度执行。操作系统快速切换线程，营造了同时执行所有线程的假象。

## 5.3 理解进程和线程优先级

使用线程，光知道创建线程或终止线程还不够，还要明白线程的有效交互离不开时间控制。时间控制有两种形式：优先级和同步。优先级控制 CPU 执行线程的频率；同步用于在共享资源时控制线程的竞争，以及安排线程执行任务的特定顺序。

当一个线程完成任务后，调度器将查找下一个要执行的线程。在选择下一个线程时，优先级较高的线程具有一定的优势。比如响应突然断电这种事情的执行优先级总是较高。执行系统中断的元素比用户进程的优先级高。其实，每个线程都有一个优先级等级表，线程基本优先级取决于它所属的进程。

线程对象特征包括基本优先级和动态优先级。用命令修改优先级只能改变基本优先级。线程优先级的修改不能超过两级，无论是向上两级还是向下两级。因为线程不可能比它的父进程还重要。虽然进程修改它的线程优先级不能超过两级，但是系统可以。对于一些要执行重要操作的线程，系统可以调整它们的优先级（即，动态优先级）。当用户在应用程序窗口执行某一动作时（如，击键或单击鼠标），系统总是会提高该窗口所属线程的优先级。当等待硬盘数据的线程获得数据时，系统也会提高该线程的优先级。这些在基本优先级的基础上临时调整的优先级，形成动态优先级。分发器或响应调度的元素，根据动态优先级选择下一个线程。基本优先级和动态优先级如图 5.1 所示。

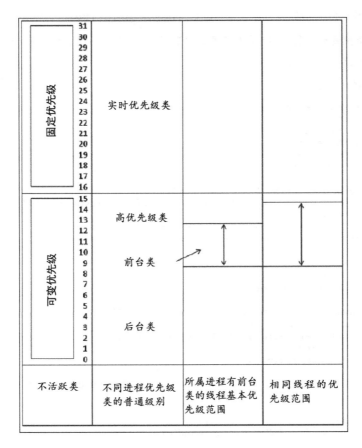

图 5.1 基本优先级和动态优先级

线程完成任务后其动态优先级立刻降低。当线程收到请求的 CPU 时间时，动态优先级每次降低一级，最终稳定在基础优先级。

下面的示例演示了进程和线程优先级的基本用法。我们将创建一个应用程序，主线程负责主窗口，其他线程负责绘制随机图案。要重点强调的是，主线程的优先级要比绘制线程的优先级至少高一级。因为主线程必须响应所有的用户输入，如移动、最小化窗口或关闭应用程序。

## 准备就绪

确定安装并运行了 Visual Studio。

## 操作步骤

现在，根据下面的步骤创建程序，稍后再详细解释。

1. 创建一个新的 C++ Win32 空项目，并命名为 basic_priority。

2. 打开【资源管理器】，右键单击【源文件】，添加一个新的源文件 basic_priority。打开 basic_priority.cpp。

3. 输入下面的代码：

```cpp
#include <windows.h>
#include <windowsx.h>
#include <time.h>
#include <tchar.h>
#include <math.h>
#include <commctrl.h>

#pragma comment ( lib, "comctl32.lib" )
#pragma comment ( linker, "\"/manifestdependency:type='win32' \
    name='Microsoft.Windows.Common-Controls' \
    version='6.0.0.0' processorArchitecture='*' \
    publicKeyToken='6595b64144ccf1df' language='*'\"" )

#define RESET_EVENT _T( "__tmp_reset_event__" )
#define WINDOW_WIDTH 800
#define WINDOW_HEIGHT 600

LRESULT CALLBACK WindowProcedure(HWND hWnd, UINT uMsg, WPARAM
    wParam, LPARAM lParam);
DWORD WINAPI StartAddress(LPVOID lpParameter);
void DrawProc(HWND hWnd);
unsigned GetNextSeed(void);

int WINAPI _tWinMain(HINSTANCE hThis, HINSTANCE hPrev, LPTSTR
    szCommandLine, int iWndShow)
{
    UNREFERENCED_PARAMETER(hPrev);
    UNREFERENCED_PARAMETER(szCommandLine);

    TCHAR* szWindowClass = _T("__priority_wnd_class__");

    WNDCLASSEX wndEx = { 0 };

    wndEx.cbSize = sizeof(WNDCLASSEX);
    wndEx.style = CS_HREDRAW | CS_VREDRAW;
    wndEx.lpfnWndProc = WindowProcedure;
    wndEx.cbClsExtra = 0;
    wndEx.cbWndExtra = 0;
    wndEx.hInstance = hThis;
    wndEx.hIcon = LoadIcon(wndEx.hInstance,
        MAKEINTRESOURCE(IDI_APPLICATION));
    wndEx.hIconSm = LoadIcon(wndEx.hInstance,
        MAKEINTRESOURCE(IDI_APPLICATION));
    wndEx.hCursor = LoadCursor(NULL, IDC_ARROW);
    wndEx.hbrBackground = (HBRUSH)(COLOR_WINDOW + 1);
    wndEx.lpszMenuName = NULL;
    wndEx.lpszClassName = szWindowClass;

    if (!RegisterClassEx(&wndEx))
    {
        return 1;
```

```cpp
    }

    HANDLE hEvent = CreateEvent(NULL, TRUE, FALSE, RESET_EVENT);

    InitCommonControls();

    HWND hWnd = CreateWindow(szWindowClass,
        _T("Basic thread priority"), WS_OVERLAPPED | WS_CAPTION
        | WS_SYSMENU | WS_MINIMIZEBOX, 50, 50, WINDOW_WIDTH,
        WINDOW_HEIGHT, NULL, NULL, wndEx.hInstance, NULL);

    if (!hWnd)
    {
        return NULL;
    }

    HFONT hFont = CreateFont(14, 0, 0, 0, FW_NORMAL, FALSE,
        FALSE, FALSE, BALTIC_CHARSET, OUT_DEFAULT_PRECIS,
        CLIP_DEFAULT_PRECIS, DEFAULT_QUALITY, DEFAULT_PITCH
        | FF_MODERN, _T("Microsoft Sans Serif"));

    ShowWindow(hWnd, iWndShow);

    MSG msg;
    while (GetMessage(&msg, 0, 0, 0))
    {
        TranslateMessage(&msg);
        DispatchMessage(&msg);
    }

    CloseHandle(hEvent);

    UnregisterClass(wndEx.lpszClassName, wndEx.hInstance);

    return (int)msg.wParam;
}

LRESULT CALLBACK WindowProcedure(HWND hWnd, UINT uMsg, WPARAM
    wParam, LPARAM lParam)
{
    switch (uMsg)
    {
        case WM_CREATE:
        {
            DWORD dwThreadId = 0;
            SetThreadPriority(GetCurrentThread(),
                THREAD_PRIORITY_NORMAL);
            HANDLE hThread = CreateThread(NULL, 0,
                (LPTHREAD_START_ROUTINE)StartAddress, hWnd,
                0, &dwThreadId); (LONG_PTR)dwThreadId;

            Sleep(100);
            CloseHandle(hThread);
            break;
        }
        case WM_CLOSE:
        {
            HANDLE hEvent = OpenEvent(EVENT_ALL_ACCESS, FALSE, RESET_EVENT);
            SetEvent(hEvent);
```

```cpp
                    DWORD dwThreadId = (DWORD)GetWindowLongPtr(hWnd, GWLP_USERDATA);
                    HANDLE hThread = OpenThread(THREAD_ALL_ACCESS, FALSE, dwThreadId);

                    WaitForSingleObject(hThread, INFINITE);
                    CloseHandle(hThread);

                    DestroyWindow(hWnd);
                    break;
                }
            case WM_DESTROY:
                {
                    PostQuitMessage(0);
                    break;
                }
            default:
                {
                    return DefWindowProc(hWnd, uMsg, wParam, lParam);
                }
        }
        return 0;
    }

DWORD WINAPI StartAddress(LPVOID lpParameter)
{
    HWND hWnd = (HWND)lpParameter;

    SetThreadPriority(GetCurrentThread(), THREAD_PRIORITY_BELOW_NORMAL);
    HANDLE hEvent = OpenEvent(EVENT_ALL_ACCESS, FALSE, RESET_EVENT);

    DWORD dwWaitResult(1);
    while (dwWaitResult != WAIT_OBJECT_0)
    {
        dwWaitResult = WaitForSingleObject(hEvent, 100);
        DrawProc(hWnd);
    }

    CloseHandle(hEvent);
    return 0L;
}

void DrawProc(HWND hWnd)
{
    int iTotal = 10;

    srand(GetNextSeed());

    HDC hDC = GetDC(hWnd);
    if (hDC)
    {
        for (int iCount = 0; iCount < iTotal; iCount++)
        {
            int iStartX = (int)(rand() % WINDOW_WIDTH);
            int iStopX = (int)(rand() % WINDOW_WIDTH);
            int iStartY = (int)(rand() % WINDOW_HEIGHT);
            int iStopY = (int)(rand() % WINDOW_HEIGHT);

            int iRed = rand() & 255;
            int iGreen = rand() & 255;
            int iBlue = rand() & 255;
```

```
            HANDLE hBrush = CreateSolidBrush(GetNearestColor(hDC,
                RGB(iRed, iGreen, iBlue)));
            HANDLE hbrOld = SelectBrush(hDC, hBrush);

            Rectangle(hDC, min(iStartX, iStopX), max(iStartX, iStopX),
                min(iStartY, iStopY), max(iStartY, iStopY));

            DeleteBrush(SelectBrush(hDC, hbrOld));
        }
        ReleaseDC(hWnd, hDC);
    }
}
unsigned GetNextSeed(void)
{
    static unsigned seed = (unsigned)time(NULL);
    return ++seed;
}
```

## 示例分析

当我们调用 CreateWindow 或 CreateWindowEx 函数时，在该函数返回之前，系统将把 WM_CREATE 消息发送给窗口消息队列。被创建窗口的窗口例程将在窗口被创建后且窗口可见之前接收到一条消息。这就是要在调用 CreateWindow 函数之前创建事件的原因。

当 WindowProcedure 接收到 WM_CREATE 消息时，我们将把主线程的优先级设置为 NORMAL（默认设置），然后再创建另一个线程负责绘制图形。等线程创建好后，通过下面的代码将其标识符与显示用户指定数据的窗口区域相关联：

```
SetWindowLongPtr(hWnd, GWLP_USERDATA, (LONG_PTR) &dwThreadId);
```

接下来，为了开始执行新创建的线程，主线程等待了 0.1 秒。然后，关闭新线程的句柄。现在，我们把线程的优先级调至比普通还低一级（THREAD_PRIORITY_BELOW_NORMAL），因为主线程必须比绘制线程的优先级至少高一级。这里要解释的是，我们假设用户希望应用程序始终都能响应。在该例中，我们的工作线程要执行的任务非常简单：绘制矩形，每绘制 10 个矩形暂停 0.1 秒。另外一种情况是，工作线程用大量的迭代计算一个**傅立叶级数**（*Fourier series*），如果计算线程的优先级等于或高于主线程的优先级，该应用程序将不会响应。这是因为分发器是根据线程的优先级和调度算法来选择执行下一个线程的。如果主线程的优先级低于工作线程，即工作线程的优先级较高，应用程序就不会立即响应用户的输入。当然，程序仍然会处理用户输入，但是要等工作线程获得 CUP 时间后才行，这将导致应用程序反应迟钝。设置线程的优先级后，绘制线程进入一个无限循环，在用户关闭应用程序之前将一直绘制矩形。

通常用正弦函数和余弦函数的无限和来表示的周期函数都能用傅立叶级数来表示。它在密码学、语音识别、手写等方面应用广泛。

现在，我们重点分析 DrawProc 绘制例程。调用 srand 函数重新开始随机生成器后，我们要获得一个设备环境来进行绘制。然后，在应用程序的窗口边界内用随机坐标绘制矩形，并在矩形中随机填色。GetNextSeed 例程把时间戳（*timestamp*）设置为种子值（作为静态变量，只在首次调用时初始化一次），这样确保了每次都是独一无二的值。然后，递增该值。

当用户要关闭应用程序时，程序将打开一个事件，并将其状态设置为 signaled。然后，线程将在下一次调用 WaitForSingleObject 时退出。我们将获得与主窗口相关联的工作线程标识符，并使用等待函数 WaitForSingleObject 等待退出。

### 更多讨论

我们也许会得出这样的结论：提高应用程序线程的优先级，应用程序执行得更快。这样想可不对。注意，如果大幅提高不同应用程序的优先级（无论是线程还是进程的优先级），系统将花费大量的时间来处理这些线程，无法执行自己的任务，最后导致系统响应慢甚至无响应。如果打开**任务管理器**，可以看到只有很少正在运行的窗口进程的优先级高于 NORMAL，通常是那些 dwm.exe（桌面窗口管理器）和 taskmgr.exe（Windows 任务管理器）。Windows 还设置了优先级更高的桌面进程以响应用户的指令，就像我们的应用程序，用户输入的优先级比绘制（计算）的优先级高。使用优先级要特别谨慎，我们的建议是，如无必要请不要提高优先级。

## 5.4 Windows 分发器对象和调度

前面介绍过，分发器从队列中选择下一个要执行的线程，从较高优先级开始到较低优先级。然而，队列中不会包含所有的线程，有些线程被挂起或被阻塞了。我们在前面的章节中介绍了线程的三种基本状态：**运行**、**就绪**和**阻塞**。现在，我们把这些状态扩展一下并解释这些状态。

在任意时刻，线程只能处于下面中的一种状态。

- **就绪**：在队列中等待执行。
- **待命**：下一个执行。
- **运行**：正在 CPU 上执行。
- **等待**：暂时不执行，正在等待恢复信号。
- **过渡**：只有当系统载入该线程的上下文才能执行。
- **终止**：执行完毕，但尚未销毁线程对象。

当分发器选择执行下一个线程时，系统把该线程的上下文载入内存。该上下文包含一系列值，如机器寄存器值、内核栈、环境块和在进程地址空间中的用户栈。虽然一部分上下文被交换到磁盘中，当线程进入过渡状态时，系统负责收集上下文的各部分。切换线程意味着要保存线程上下文中的所有内容，并载入下一个线程的所有上下文。

新载入的线程执行一段时间，这段时间叫做**时间片**（*quantum*）。系统维护一个测量当前时间的计数器。对于每个时钟周期，系统以某值递减计数器。当计数器为零时，分发器执行上下文切换，准备执行下一个线程。

## 5.5 使用互斥量

互斥量是一个 Windows 同步对象。要对共享对象实现独占访问就要用到互斥量，需要保护某部分代码或一次执行一个线程时也可以使用互斥量。

在下面的示例中，我们要求用户提供一个每行有 6 个整数坐标的文本文件，用于计算一个包含两个线性方程的方程组：

$$\begin{cases} a_1x+b_1y=c_1 \\ a_2x+b_2y=c_2 \end{cases}$$

示例要用到第 1 章中实现的 CQueue 类。我们的任务是载入文本文件中的所有值，然后把每组坐标添加到队列中。每载入一组坐标，就启动一些线程执行并发计算。这里，最重要的是确定应该启动多少个线程，以及保护和同步哪部分代码。

### 准备就绪

确定安装并运行了 Visual Studio。

### 操作步骤

现在，根据下面的步骤创建程序，稍后再详细解释。

1. 创建一个新的 C++ Win32 项目，并命名为 concurrent_operations。把第 1 章中的头文件 CQueue.h 复制到该项目的目录中。

2. 打开【解决方案资源管理器】，右键单击【头文件】，选择【添加】-【现有项】，添加 CQueue.h。

3. 打开【解决方案资源管理器】，右键单击【头文件】，添加新的头文件 main。打开 main.h。

4. 输入下面的代码：

   ```
   #pragma once
   ```

```cpp
#include <windows.h>
#include <commctrl.h>
#include <tchar.h>
#include <psapi.h>
#include <strsafe.h>
#include <string.h>
#include <sstream>
#include <fstream>
#include <commdlg.h>
#include "CQueue.h"

#pragma comment ( lib, "comctl32.lib" )
#pragma comment ( linker, "\"/manifestdependency:type='win32' \
name='Microsoft.Windows.Common-Controls' \
version='6.0.0.0' processorArchitecture='*' \
publicKeyToken='6595b64144ccf1df' language='*'\"" )

using namespace std;

#define CONTROL_BROWSE 100
#define CONTROL_START 101
#define CONTROL_RESULT 103
#define CONTROL_TEXT 104
#define CONTROL_PROGRESS 105

LRESULT CALLBACK WindowProcedure(HWND hWnd, UINT uMsg, WPARAM
    wParam, LPARAM lParam);
DWORD WINAPI StartAddress(LPVOID lpParameter);
BOOL FileDialog(HWND hWnd, LPSTR szFileName);

typedef struct
{
    int iA1;
    int iB1;
    int iC1;
    int iA2;
    int iB2;
    int iC2;
    HWND hWndProgress;
    HWND hWndResult;
} QueueElement, *PQueueElement;
```

5. 打开【解决方案资源管理器】，右键单击【源文件】，添加一个新的源文件 main。打开 main.cpp。

6. 输入下面的代码：

```cpp
#pragma once

/************************************************************
* 用户指定包含坐标的文本文件 *
* 本例将计算有两个线性方程的方程组*
* a1x + b1y = c1; *
* a2x + b2y = c2; *
* *
* 文本中每行有6个坐标 *
* 格式如下： *
* *
* 4 -3 2 -7 5 6 *
```

```
 *  12  1  -7  9  -5 8  *
 *  *
 *  两行数据不要用空行隔开  *
 **********************************************/

#include "main.h"

char szFilePath[MAX_PATH] = { 0 };
char szResult[4096];
CQueue<QueueElement> queue;
TCHAR* szMutex = _T("__mutex_132__");

int WINAPI _tWinMain(HINSTANCE hThis, HINSTANCE hPrev,
    LPTSTR szCommandLine, int iWndShow)
{
    UNREFERENCED_PARAMETER(hPrev);
    UNREFERENCED_PARAMETER(szCommandLine);

    TCHAR* szWindowClass = _T("__concurrent_operations__");
    WNDCLASSEX wndEx = { 0 };

    wndEx.cbSize = sizeof(WNDCLASSEX);
    wndEx.style = CS_HREDRAW | CS_VREDRAW;
    wndEx.lpfnWndProc = WindowProcedure;
    wndEx.cbClsExtra = 0;
    wndEx.cbWndExtra = 0;
    wndEx.hInstance = hThis;
    wndEx.hIcon = LoadIcon(wndEx.hInstance, MAKEINTRESOURCE(IDI_APPLICATION));
    wndEx.hCursor = LoadCursor(NULL, IDC_ARROW);
    wndEx.hbrBackground = (HBRUSH)(COLOR_WINDOW + 1);
    wndEx.lpszMenuName = NULL;
    wndEx.lpszClassName = szWindowClass;
    wndEx.hIconSm = LoadIcon(wndEx.hInstance, MAKEINTRESOURCE(IDI_APPLICATION));

    if (!RegisterClassEx(&wndEx))
    {
        return 1;
    }

    InitCommonControls();

    HWND hWnd = CreateWindow(szWindowClass,
        _T("Concurrent operations"), WS_OVERLAPPED
        | WS_CAPTION | WS_SYSMENU, 50, 50, 305, 250, NULL, NULL,
        wndEx.hInstance, NULL);

    if (!hWnd)
    {
        return NULL;
    }

    HFONT hFont = CreateFont(14, 0, 0, 0, FW_NORMAL, FALSE,
        FALSE, FALSE, BALTIC_CHARSET, OUT_DEFAULT_PRECIS,
        CLIP_DEFAULT_PRECIS, DEFAULT_QUALITY, DEFAULT_PITCH
        | FF_MODERN, _T("Microsoft Sans Serif"));

    HWND hLabel = CreateWindow(_T("STATIC"),
        _T(" Press \"Browse\" and choose file with coordinates: "),
        WS_CHILD | WS_VISIBLE | SS_CENTERIMAGE | SS_LEFT | WS_BORDER,
```

```
            20, 20, 250, 25, hWnd, (HMENU)CONTROL_TEXT, wndEx.hInstance, NULL);
        SendMessage(hLabel, WM_SETFONT, (WPARAM)hFont, TRUE);

        HWND hResult = CreateWindow(_T("STATIC"), _T(""), WS_CHILD |
            WS_VISIBLE | SS_LEFT | WS_BORDER, 20, 65, 150, 75, hWnd,
            (HMENU)CONTROL_RESULT, wndEx.hInstance, NULL);
        SendMessage(hResult, WM_SETFONT, (WPARAM)hFont, TRUE);

        HWND hBrowse = CreateWindow(_T("BUTTON"), _T("Browse"), WS_CHILD
            | WS_VISIBLE
            | BS_PUSHBUTTON, 190, 65, 80, 25, hWnd, (HMENU)CONTROL_BROWSE,
            wndEx.hInstance, NULL);
        SendMessage(hBrowse, WM_SETFONT, (WPARAM)hFont, TRUE);

        HWND hStart = CreateWindow(_T("BUTTON"), _T("Start"), WS_CHILD |
            WS_VISIBLE
            | BS_PUSHBUTTON, 190, 115, 80, 25, hWnd, (HMENU)CONTROL_START,
            wndEx.hInstance, NULL);
        SendMessage(hStart, WM_SETFONT, (WPARAM)hFont, TRUE);

        HWND hProgress = CreateWindow(PROGRESS_CLASS, _T(""), WS_CHILD |
            WS_VISIBLE
            | WS_BORDER, 20, 165, 250, 25, hWnd, (HMENU)CONTROL_PROGRESS,
            wndEx.hInstance, NULL);
        SendMessage(hProgress, PBM_SETSTEP, (WPARAM)1, 0);
        SendMessage(hProgress, PBM_SETPOS, (WPARAM)0, 0);

        ShowWindow(hWnd, iWndShow);

        HANDLE hMutex = CreateMutex(NULL, FALSE, szMutex);

        MSG msg;
        while (GetMessage(&msg, 0, 0, 0))
        {
            TranslateMessage(&msg);
            DispatchMessage(&msg);
        }

        CloseHandle(hMutex);
        UnregisterClass(wndEx.lpszClassName, wndEx.hInstance);

        return (int)msg.wParam;
    }

    LRESULT CALLBACK WindowProcedure(HWND hWnd, UINT uMsg, WPARAM
        wParam, LPARAM lParam)
    {
        switch (uMsg)
        {
            case WM_DESTROY:
            {
                PostQuitMessage(0);
                break;
            }
            case WM_COMMAND:
            {
                switch (LOWORD(wParam))
                {
                    case CONTROL_BROWSE:
```

```cpp
        {
            if (!FileDialog(hWnd, szFilePath))
            {
                MessageBox(hWnd,
                    _T("You must choose valid file path!"),
                    _T("Error!"), MB_OK | MB_TOPMOST | MB_ICONERROR);
            }
            else
            {
                char szBuffer[MAX_PATH];
                wsprintfA(szBuffer,
                    "\n File: %s Press \"Start\" now.",
                    strrchr(szFilePath, '\\') + 1);

                SetWindowTextA(GetDlgItem(hWnd, CONTROL_TEXT),
                    szBuffer);
            }
            break;
        }
        case CONTROL_START:
        {
            if (!*szFilePath)
            {
                MessageBox(hWnd,
                    _T("You must choose valid file path first!"),
                    _T("Error!"), MB_OK | MB_TOPMOST | MB_ICONERROR);
                break;
            }
            ifstream infile(szFilePath);

            if (infile.is_open())
            {
                string line;
                while (std::getline(infile, line))
                {
                    QueueElement* pQElement = new QueueElement();
                    istringstream iss(line);

                    if (!(iss >> pQElement->iA1 >> pQElement->iB1 >>
                        pQElement->iC1 >> pQElement->iA2 >>
                        pQElement->iB2 >> pQElement->iC2))
                    {
                        break;
                    }

                    pQElement->hWndProgress = GetDlgItem(hWnd,
                        CONTROL_PROGRESS);
                    pQElement->hWndResult = GetDlgItem(hWnd,
                        CONTROL_RESULT);

                    queue.Enqueue(pQElement);
                }
                infile.close();

                SendMessage(GetDlgItem(hWnd, CONTROL_PROGRESS),
                    PBM_SETRANGE, 0, MAKELPARAM(0, queue.Count()));

                SYSTEM_INFO sysInfo;
                GetSystemInfo(&sysInfo);
```

```
                                for (DWORD dwIndex = 0; dwIndex <
                                    sysInfo.dwNumberOfProcessors; dwIndex++)
                                {
                                    HANDLE hThread = CreateThread(NULL, 0,
                                        (LPTHREAD_START_ROUTINE)StartAddress,&queue,
                                        0, NULL);
                                    SetThreadIdealProcessor(hThread, dwIndex);
                                    Sleep(100);
                                    CloseHandle(hThread);
                                }
                            }
                            else
                            {
                                MessageBox(hWnd, _T("Cannot open file!"),
                                    _T("Error!"), MB_OK | MB_TOPMOST | MB_ICONERROR);
                            }
                            break;
                        }
                    }
                    break;
                }
                default:
                {
                    return DefWindowProc(hWnd, uMsg, wParam, lParam);
                }
            }
            return 0;
        }

        DWORD WINAPI StartAddress(LPVOID lpParameter)
        {
            // 用克莱姆法则解线性方程组
            CQueue<QueueElement>* pQueue = (CQueue<QueueElement>*)lpParameter;

            HANDLE hMutex = OpenMutex(MUTEX_ALL_ACCESS, FALSE, szMutex);
            QueueElement* pQElement = NULL;

            while (true)
            {
                WaitForSingleObject(hMutex, INFINITE);

                pQElement = pQueue->Dequeue();

                ReleaseMutex(hMutex);

                if (pQElement == NULL)
                {
                    break;
                }

                char szBuffer[1024];

                double dDeterminant = (pQElement->iA1 * pQElement->iB2) -
                    (pQElement->iB1 * pQElement->iA2);

                if (dDeterminant != 0)
                {
                    double dX = ((pQElement->iC1 * pQElement->iB2) -
                        (pQElement->iB1 * pQElement->iC2)) / dDeterminant;
```

```
                double dY = ((pQElement->iA1 * pQElement->iC2) -
                    (pQElement->iC1 * pQElement->iA2)) / dDeterminant;
                sprintf_s(szBuffer, " x = %8.4lf,\ty = %8.4lf\n", dX, dY);
            }
            else
            {
                sprintf_s(szBuffer, " Determinant is zero.\n");
            }

            strcat_s(szResult, 4096, szBuffer);
            SetWindowTextA(pQElement->hWndResult, szResult);

            SendMessage(pQElement->hWndProgress, PBM_STEPIT, 0, 0);
            delete pQElement;

            Sleep(1000);
        }

        CloseHandle(hMutex);
        return 0L;
    }

    BOOL FileDialog(HWND hWnd, LPSTR szFileName)
    {
        OPENFILENAMEA ofn;

        char szFile[MAX_PATH];

        ZeroMemory(&ofn, sizeof(ofn));

        ofn.lStructSize = sizeof(ofn);
        ofn.hwndOwner = hWnd;
        ofn.lpstrFile = szFile;
        ofn.lpstrFile[0] = '\0';
        ofn.nMaxFile = sizeof(szFile);
        ofn.lpstrFilter = "All\0*.*\0Text\0*.TXT\0";
        ofn.nFilterIndex = 1;
        ofn.lpstrFileTitle = NULL;
        ofn.nMaxFileTitle = 0;
        ofn.lpstrInitialDir = NULL;
        ofn.Flags = OFN_PATHMUSTEXIST | OFN_FILEMUSTEXIST;

        if (GetOpenFileNameA(&ofn) == TRUE)
        {
            strcpy_s(szFileName, MAX_PATH - 1, szFile);
            return TRUE;
        }

        return FALSE;
    }
```

## 示例分析

为了更容易地把每一组坐标载入队列,我们在 main.h 头文件中定义了 QueueElement 结构。在 main.cpp 中,声明了一些全局变量:szFilePath(表示用户要选择的文件)、szResult(根据计算结果显示消息)、queue(表示实际的队列)、szMutex(表示稍后同步的互斥量名)。在应用程序的入口

(_tWinMain)，初始化并创建了带有多个控件的窗口。hLabel 控件用来显示与用户交互的消息；hResult 控件，用户用来选择文件；hStart 控件，用户用来开始计算；hProgress 控件，应用程序用来显示计算进度。

主线程在进入消息循环之前，将创建一个互斥量对象。当用户单击 **Browse** 按钮时，程序将显示 **Open file** 对话框，供用户选择合适的文件。当用户单击 **Start** 按钮时，程序将检查用户所选的文件是否是之前选定的。如果是，程序将从文件中每次读取一行，并初始化 QueueElement 对象。QueueElement 对象需要坐标值，并提供 progress 控件的句柄和 result 控件的句柄。线程将使用这些句柄分别显示计算进度和结果。

我们在前面曾多次提到过，应用程序设计得当能减少出错和漏洞。首先要注意的问题是，要启动多少个线程？下一个问题是同步访问中要保护哪部分的代码？或者要让哪些语句获得独占访问？第一个问题的答案是：如果执行的并发任务比可用的处理器数量少，应该尽量多启动线程。然而，用户可以提供一个文件中有无限数量行的坐标。如果要执行的任务比可用处理器数量多，系统就不会执行得很好，应用程序也发挥不出最佳的性能。因此，当把所有的对象都载入队列时，要询问系统有多少可用的 CPU。然后，限制线程的数量不超过可用处理器的数量，代码如下所示：

```
SYSTEM_INFO sysInfo;
GetSystemInfo(&sysInfo);

for (DWORD dwIndex = 0; dwIndex < sysInfo.dwNumberOfProcessors; dwIndex++)
{
    HANDLE hThread = CreateThread(NULL, 0,(LPTHREAD_START_ROUTINE)
        StartAddress, &queue, 0, NULL);
    SetThreadIdealProcessor(hThread, dwIndex);
    Sleep( 100 );
    CloseHandle( hThread );
}
```

除此之外，还要询问队列计数，防止启动多余的线程，但前提是你要意识到这一点。我们用 SetThreadIdealProcessor API 来通知系统把线程调度给已选的处理器。该例程指定一个线程的理想处理器，并给调度器提供更适合线程的处理器线索。调度器尽可能在理想的处理器上运行线程，但是无法保证一定是例程指定的处理器。如果所选的处理器正忙，那么下一次才能执行线程，系统会把线程调度到下一个可用的内核上。

第二个问题的答案是：所有线程要使用的共享对象是什么？肯定是 queue 对象。每个线程都并行执行 CQueue::Dequeue 操作。没有合适的同步，这就是应用程序的瓶颈，必须要特别注意。我们使用互斥量对象来独占访问线程。这正是 Dequeue 操作需要的，因为在任意时间，应该只有一个线程从队列中获得一个对象。使用互斥量对象，就能为共享对象提供独占访问。在本例中，共享对象是 queue 和它的操作 Dequeue。

### 更多讨论

这个示例比较简单，仅为了演示如何建立从设计到实现应用程序的方法。虽然第 7 章将详细介绍如何编写并发代码,但是熟悉并发执行和养成在编写代码前认真思考的好习惯非常重要。

## 5.6 使用信号量

信号量也是 Windows 同步对象。信号量类似一个窄门,在任意时间只允许一定数量的线程通过。我们将利用上一个示例,仅作一些小改动,演示如何用不同的同步对象和技术解决相同的问题。

### 准备就绪

确定安装并运行了 Visual Studio。

### 操作步骤

现在,根据下面的步骤创建程序,稍后再详细解释。

1. 创建一个新的 C++ Win32 空项目,并命名为 concurrent_operations2。把第 1 章中的头文件 CQueue.h 复制到该项目的目录中。

2. 打开【解决方案资源管理器】,右键单击【头文件】,选择【添加】-【现有项】,添加 CQueue.h。

3. 打开【解决方案资源管理器】,右键单击【头文件】,添加新的头文件 main。打开 main.h。

4. 输入下面的代码:

   ```
   #pragma once

   #include <windows.h>
   #include <commctrl.h>
   #include <tchar.h>
   #include <psapi.h>
   #include <strsafe.h>
   #include <string.h>
   #include <sstream>
   #include <fstream>
   #include <commdlg.h>
   #include "CQueue.h"

   #pragma comment ( lib, "comctl32.lib" )
   #pragma comment ( linker, "\"/manifestdependency:type='win32' \
   name='Microsoft.Windows.Common-Controls' \
   version='6.0.0.0' processorArchitecture='*' \
   publicKeyToken='6595b64144ccf1df' language='*'\"" )

   using namespace std;

   #define CONTROL_BROWSE 100
   #define CONTROL_START 101
   #define CONTROL_RESULT 103
   #define CONTROL_TEXT 104
   ```

```
#define CONTROL_PROGRESS 105

LRESULT CALLBACK WindowProcedure(HWND hWnd, UINT uMsg, WPARAM
    wParam, LPARAM lParam);
DWORD WINAPI StartAddress(LPVOID lpParameter);
BOOL FileDialog(HWND hWnd, LPSTR szFileName);

typedef struct
{
    int iA1;
    int iB1;
    int iC1;
    int iA2;
    int iB2;
    int iC2;
    HWND hWndProgress;
    HWND hWndResult;
} QueueElement, *PQueueElement;
```

5. 打开【解决方案资源管理器】，右键单击【源文件】，添加一个新的源文件 main。打开 main.cpp。

6. 输入下面的代码：

```
#pragma once

/***********************************************************
* 用户指定包含坐标的文本文件 *
* 本例将计算有两个线性方程的方程组 *
* a1x + b1y = c1; *
* a2x + b2y = c2; *
* *
* 文本中每行有6个坐标 *
* 格式如下： *
* *
* 4 -3 2 -7 5 6 *
* 12 1 -7 9 -5 8 *
* *
* 两行数据不要用空行隔开 *
***********************************************************/

#include "main.h"

char szFilePath[MAX_PATH] = { 0 };
char szResult[4096];
CQueue<QueueElement> queue;
TCHAR* szMutex = _T("__mutex_165__");
TCHAR* szSemaphore = _T("__semaphore_456__");

int WINAPI _tWinMain(HINSTANCE hThis, HINSTANCE hPrev, LPTSTR
    szCommandLine, int iWndShow)
{
    UNREFERENCED_PARAMETER(hPrev);
    UNREFERENCED_PARAMETER(szCommandLine);

    TCHAR* szWindowClass = _T("__concurrent_operations__");
    WNDCLASSEX wndEx = { 0 };
```

```
wndEx.cbSize = sizeof(WNDCLASSEX);
wndEx.style = CS_HREDRAW | CS_VREDRAW;
wndEx.lpfnWndProc = WindowProcedure;
wndEx.cbClsExtra = 0;
wndEx.cbWndExtra = 0;
wndEx.hInstance = hThis;
wndEx.hIcon = LoadIcon(wndEx.hInstance,
    MAKEINTRESOURCE(IDI_APPLICATION));
wndEx.hCursor = LoadCursor(NULL, IDC_ARROW);
wndEx.hbrBackground = (HBRUSH)(COLOR_WINDOW + 1);
wndEx.lpszMenuName = NULL;
wndEx.lpszClassName = szWindowClass;
wndEx.hIconSm = LoadIcon(wndEx.hInstance,
    MAKEINTRESOURCE(IDI_APPLICATION));

if (!RegisterClassEx(&wndEx))
{
    return 1;
}

InitCommonControls();

HWND hWnd = CreateWindow(szWindowClass,
    _T("Concurrent operations"), WS_OVERLAPPED
    | WS_CAPTION | WS_SYSMENU, 50, 50, 305, 250, NULL, NULL,
    wndEx.hInstance, NULL);

if (!hWnd)
{
    return NULL;
}

HFONT hFont = CreateFont(14, 0, 0, 0, FW_NORMAL, FALSE,
    FALSE, FALSE, BALTIC_CHARSET, OUT_DEFAULT_PRECIS,
    CLIP_DEFAULT_PRECIS, DEFAULT_QUALITY, DEFAULT_PITCH
    | FF_MODERN, _T("Microsoft Sans Serif"));

HWND hLabel = CreateWindow(_T("STATIC"),
    _T(" Press \"Browse\" and choose file with coordinates: "),
    WS_CHILD | WS_VISIBLE | SS_CENTERIMAGE | SS_LEFT | WS_BORDER,
    20, 20, 250, 25, hWnd, (HMENU)CONTROL_TEXT, wndEx.hInstance, ULL);

SendMessage(hLabel, WM_SETFONT, (WPARAM)hFont, TRUE);

HWND hResult = CreateWindow(_T("STATIC"), _T(""), WS_CHILD |
    WS_VISIBLE | SS_LEFT | WS_BORDER, 20, 65, 150, 75, hWnd,
    (HMENU)CONTROL_RESULT, wndEx.hInstance, NULL);
SendMessage(hResult, WM_SETFONT, (WPARAM)hFont, TRUE);

HWND hBrowse = CreateWindow(_T("BUTTON"), _T("Browse"), WS_CHILD
    | WS_VISIBLE
    | BS_PUSHBUTTON, 190, 65, 80, 25, hWnd, (HMENU)CONTROL_BROWSE,
    wndEx.hInstance, NULL);
SendMessage(hBrowse, WM_SETFONT, (WPARAM)hFont, TRUE);

HWND hStart = CreateWindow(_T("BUTTON"), _T("Start"), WS_CHILD |
    WS_VISIBLE
    | BS_PUSHBUTTON, 190, 115, 80, 25, hWnd, (HMENU)CONTROL_START,
    wndEx.hInstance, NULL);
```

```
    SendMessage(hStart, WM_SETFONT, (WPARAM)hFont, TRUE);

    HWND hProgress = CreateWindow(PROGRESS_CLASS, _T(""), WS_CHILD |
        WS_VISIBLE
        | WS_BORDER, 20, 165, 250, 25, hWnd, (HMENU)CONTROL_PROGRESS,
        wndEx.hInstance, NULL);
    SendMessage(hProgress, PBM_SETSTEP, (WPARAM)1, 0);
    SendMessage(hProgress, PBM_SETPOS, (WPARAM)0, 0);

    ShowWindow(hWnd, iWndShow);

    HANDLE hMutex = CreateMutex(NULL, FALSE, szMutex);

    SYSTEM_INFO sysInfo;
    GetSystemInfo(&sysInfo);
    LONG lProcessorCount = (LONG)sysInfo.dwNumberOfProcessors;
    HANDLE hSemaphore = CreateSemaphore(NULL, lProcessorCount,
        lProcessorCount, szSemaphore);

    MSG msg;
    while (GetMessage(&msg, 0, 0, 0))
    {
        TranslateMessage(&msg);
        DispatchMessage(&msg);
    }

    CloseHandle(hSemaphore);
    CloseHandle(hMutex);
    UnregisterClass(wndEx.lpszClassName, wndEx.hInstance);

    return (int)msg.wParam;
}

LRESULT CALLBACK WindowProcedure(HWND hWnd, UINT uMsg, WPARAM
    wParam, LPARAM lParam)
{
    switch (uMsg)
    {
        case WM_DESTROY:
        {
            PostQuitMessage(0);
            break;
        }
        case WM_COMMAND:
        {
            switch (LOWORD(wParam))
            {
                case CONTROL_BROWSE:
                {
                    if (!FileDialog(hWnd, szFilePath))
                    {
                        MessageBox(hWnd,
                            _T("You must choose valid file path!"),
                            _T("Error!"), MB_OK | MB_TOPMOST | MB_ICONERROR);
                    }
                    else
                    {
                        char szBuffer[MAX_PATH];
                        wsprintfA(szBuffer,
```

```cpp
                        "\n File: %s Press \"Start\" now.",
                        strrchr(szFilePath, '\\') + 1);
                    SetWindowTextA(GetDlgItem(hWnd, CONTROL_TEXT),
                        szBuffer);
            }
            break;
        }
        case CONTROL_START:
        {
            if (!*szFilePath)
            {
                MessageBox(hWnd,
                    _T("You must choose valid file path first!"),
                    _T("Error!"), MB_OK | MB_TOPMOST | MB_ICONERROR);
                break;
            }

            ifstream infile(szFilePath);

            if (infile.is_open())
            {
                string line;
                while (std::getline(infile, line))
                {
                    QueueElement* pQElement = new QueueElement();

                    istringstream iss(line);

                    if (!(iss >> pQElement->iA1 >> pQElement->iB1 >>
                        pQElement->iC1 >> pQElement->iA2 >>
                        pQElement->iB2 >> pQElement->iC2))
                    {
                        break;
                    }

                    pQElement->hWndProgress = GetDlgItem(hWnd,
                        CONTROL_PROGRESS);
                    pQElement->hWndResult = GetDlgItem(hWnd,
                        CONTROL_RESULT);

                    queue.Enqueue(pQElement);
                }

                infile.close();

                SendMessage(GetDlgItem(hWnd, CONTROL_PROGRESS),
                    PBM_SETRANGE, 0, MAKELPARAM(0, queue.Count()));

                int iCount = queue.Count();

                for (int iIndex = 0; iIndex < iCount; iIndex++)
                {
                    HANDLE hThread = CreateThread(NULL, 0,
                        (LPTHREAD_START_ROUTINE)StartAddress,
                        &queue, 0, NULL);
                    Sleep(100);
                    CloseHandle(hThread);
                }
            }
```

```cpp
                    else
                    {
                        MessageBox(hWnd, _T("Cannot open file!"),
                            _T("Error!"), MB_OK | MB_TOPMOST | MB_ICONERROR);
                    }
                    break;
                }
            }
            break;
        }
        default:
        {
            return DefWindowProc(hWnd, uMsg, wParam, lParam);
        }
    }
    return 0;
}

DWORD WINAPI StartAddress(LPVOID lpParameter)
{
    // 用克莱姆法则解线性方程组
    CQueue<QueueElement>* pQueue = (CQueue<QueueElement>*)
        lpParameter;

    HANDLE hMutex = OpenMutex(MUTEX_ALL_ACCESS, FALSE, szMutex);
    HANDLE hSemaphore = OpenSemaphore(SEMAPHORE_ALL_ACCESS, FALSE,
        szSemaphore);
    QueueElement* pQElement = NULL;

    while (true)
    {
        WaitForSingleObject(hSemaphore, INFINITE);
        ReleaseSemaphore(hSemaphore, 1, NULL);

        WaitForSingleObject(hMutex, INFINITE);
        pQElement = pQueue->Dequeue();
        ReleaseMutex(hMutex);

        if (pQElement == NULL)
        {
            break;
        }

        char szBuffer[1024];

        double dDeterminant = (pQElement->iA1 * pQElement->iB2) -
            (pQElement->iB1 * pQElement->iA2);

        if (dDeterminant != 0)
        {
            double dX = ((pQElement->iC1 * pQElement->iB2) -
                (pQElement->iB1 * pQElement->iC2)) / dDeterminant;
            double dY = ((pQElement->iA1 * pQElement->iC2) -
                (pQElement->iC1 * pQElement->iA2)) / dDeterminant;
            sprintf_s(szBuffer, " x = %8.4lf,\ty = %8.4lf\n", dX, dY);
        }
        else
        {
            sprintf_s(szBuffer, " Determinant is zero.\n");
```

```
            }

            strcat_s(szResult, 4096, szBuffer);
            SetWindowTextA(pQElement->hWndResult, szResult);

            SendMessage(pQElement->hWndProgress, PBM_STEPIT, 0, 0);

            delete pQElement;

            Sleep(1000);
        }

        CloseHandle(hSemaphore);
        CloseHandle(hMutex);

        return 0L;
    }
    BOOL FileDialog(HWND hWnd, LPSTR szFileName)
    {
        OPENFILENAMEA ofn;

        char szFile[MAX_PATH];

        ZeroMemory(&ofn, sizeof(ofn));

        ofn.lStructSize = sizeof(ofn);
        ofn.hwndOwner = hWnd;
        ofn.lpstrFile = szFile;
        ofn.lpstrFile[0] = '\0';
        ofn.nMaxFile = sizeof(szFile);
        ofn.lpstrFilter = "All\0*.*\0Text\0*.TXT\0";
        ofn.nFilterIndex = 1;
        ofn.lpstrFileTitle = NULL;
        ofn.nMaxFileTitle = 0;
        ofn.lpstrInitialDir = NULL;
        ofn.Flags = OFN_PATHMUSTEXIST | OFN_FILEMUSTEXIST;

        if (GetOpenFileNameA(&ofn) == TRUE)
        {
            strcpy_s(szFileName, MAX_PATH - 1, szFile);
            return TRUE;
        }

        return FALSE;
    }
```

### 示例分析

该例与 5.5 节的示例稍有不同,没有创建和可用处理器数量一样多的线程,而是实例化了一个信号量对象,这样并行操作的数量就不会超过 CPU 的数量。与此类似,两个示例唯一的区别是所选的方法不同。本例在 StartAddress 例程中向系统询问信号量对象。成功操作后,通过递减信号量的数量释放信号量,如下所示:

```
WaitForSingleObject( hSemaphore, INFINITE );
ReleaseSemaphore( hSemaphore, 1, NULL );
```

## 更多讨论

再次强调,最重要的部分是合理选择同步部分和同步对象。我们故意用信号量,是为了让读者明白一个问题有多种解决方案。还好,Windows 提供了多种同步对象,开发人员应该了解各同步对象之间的异同,方便在开发过程中选择最合适的对象。

## 5.7 使用事件

事件同步对象的工作方式类似以某种形式警告线程的信号。它可在用户进行输入时,或者用于开始或停止执行一个线程,等等。

我们还是用相同的示例,然后添加一个事件来改善应用程序的行为,该事件将触发那些用户要中止执行并结束程序的线程。用这种方法,应用程序对用户更加友好,而且能更积极地响应用户的输入。

### 准备就绪

确定安装并运行了 Visual Studio。

### 操作步骤

现在,根据下面的步骤创建程序,稍后再详细解释。

1. 创建一个新的 C++ Win32 空项目,并命名为 concurrent_operations3。把第 1 章中的头文件 CQueue.h 复制到该项目的目录中。

2. 打开【解决方案资源管理器】,右键单击【头文件】,选择【添加】-【现有项】,添加 CQueue.h。

3. 打开【解决方案资源管理器】,右键单击【头文件】,添加新的头文件 main。打开 main.h。

4. 输入下面的代码:

```
#pragma once

#include <windows.h>
#include <commctrl.h>
#include <tchar.h>
#include <psapi.h>
#include <strsafe.h>
#include <string.h>
#include <sstream>
#include <fstream>
#include <commdlg.h>
#include "CQueue.h"

#pragma comment ( lib, "comctl32.lib" )
```

```
#pragma comment ( linker, "\"/manifestdependency:type='win32' \
name='Microsoft.Windows.Common-Controls' \
version='6.0.0.0' processorArchitecture='*' \
publicKeyToken='6595b64144ccf1df' language='*'\"" )

using namespace std;

#define CONTROL_BROWSE 100
#define CONTROL_START 101
#define CONTROL_RESULT 103
#define CONTROL_TEXT 104
#define CONTROL_PROGRESS 105

LRESULT CALLBACK WindowProcedure(HWND hWnd, UINT uMsg,
    WPARAM wParam, LPARAM lParam);
DWORD WINAPI StartAddress(LPVOID lpParameter);
BOOL FileDialog(HWND hWnd, LPSTR szFileName);

typedef struct
{
    int iA1;
    int iB1;
    int iC1;
    int iA2;
    int iB2;
    int iC2;
    HWND hWndProgress;
    HWND hWndResult;
} QueueElement, *PQueueElement;
```

5. 打开【解决方案资源管理器】，右键单击【源文件】，添加一个新的源文件 main。打开 main.cpp。

6. 输入下面的代码：

```
#pragma once

/***********************************************************
* 用户指定包含坐标的文本文件 *
* 本例将计算有两个线性方程的方程组 *
* a1x + b1y = c1; *
* a2x + b2y = c2; *
* *
* 文本中每行有 6 个坐标 *
* 格式如下： *
* *
* 4 -3 2 -7 5 6 *
* 12 1 -7 9 -5 8 *
* *
* 两行数据不要用空行隔开 *
***********************************************************/

#include "main.h"

char szFilePath[MAX_PATH] = { 0 };
char szResult[4096];
CQueue<QueueElement> queue;
TCHAR* szEvent = _T("__event_879__");
```

```c
TCHAR* szMutex = _T("__mutex_132__");

int WINAPI _tWinMain(HINSTANCE hThis, HINSTANCE hPrev, LPTSTR
    szCommandLine, int iWndShow)
{
    UNREFERENCED_PARAMETER(hPrev);
    UNREFERENCED_PARAMETER(szCommandLine);

    TCHAR* szWindowClass = _T("__concurrent_operations__");

    WNDCLASSEX wndEx = { 0 };

    wndEx.cbSize = sizeof(WNDCLASSEX);
    wndEx.style = CS_HREDRAW | CS_VREDRAW;
    wndEx.lpfnWndProc = WindowProcedure;
    wndEx.cbClsExtra = 0;
    wndEx.cbWndExtra = 0;
    wndEx.hInstance = hThis;
    wndEx.hIcon = LoadIcon(wndEx.hInstance, MAKEINTRESOURCE(IDI_APPLICATION));
    wndEx.hCursor = LoadCursor(NULL, IDC_ARROW);
    wndEx.hbrBackground = (HBRUSH)(COLOR_WINDOW + 1);
    wndEx.lpszMenuName = NULL;
    wndEx.lpszClassName = szWindowClass;
    wndEx.hIconSm = LoadIcon(wndEx.hInstance, MAKEINTRESOURCE(IDI_APPLICATION));

    if (!RegisterClassEx(&wndEx))
    {
        return 1;
    }

    InitCommonControls();

    HWND hWnd = CreateWindow(szWindowClass,
        _T("Concurrent operations"), WS_OVERLAPPED
        | WS_CAPTION | WS_SYSMENU, 50, 50, 305, 250, NULL, NULL,
        wndEx.hInstance, NULL);

    if (!hWnd)
    {
        return NULL;
    }

    HFONT hFont = CreateFont(14, 0, 0, 0, FW_NORMAL, FALSE,
        FALSE, FALSE, BALTIC_CHARSET, OUT_DEFAULT_PRECIS,
        CLIP_DEFAULT_PRECIS, DEFAULT_QUALITY, DEFAULT_PITCH
        | FF_MODERN, _T("Microsoft Sans Serif"));

    HWND hLabel = CreateWindow(_T("STATIC"),
        _T(" Press \"Browse\" and choose file with coordinates: "),
        WS_CHILD | WS_VISIBLE | SS_CENTERIMAGE | SS_LEFT | WS_BORDER,
        20, 20, 250, 25, hWnd, (HMENU)CONTROL_TEXT, wndEx.hInstance, NULL);

    SendMessage(hLabel, WM_SETFONT, (WPARAM)hFont, TRUE);

    HWND hResult = CreateWindow(_T("STATIC"), _T(""), WS_CHILD |
        WS_VISIBLE | SS_LEFT | WS_BORDER, 20, 65, 150, 75, hWnd,
        (HMENU)CONTROL_RESULT, wndEx.hInstance, NULL);
    SendMessage(hResult, WM_SETFONT, (WPARAM)hFont, TRUE);
```

```
    HWND hBrowse = CreateWindow(_T("BUTTON"), _T("Browse"), WS_CHILD
        | WS_VISIBLE
        | BS_PUSHBUTTON, 190, 65, 80, 25, hWnd, (HMENU)CONTROL_BROWSE,
        wndEx.hInstance, NULL);
    SendMessage(hBrowse, WM_SETFONT, (WPARAM)hFont, TRUE);

    HWND hStart = CreateWindow(_T("BUTTON"), _T("Start"), WS_CHILD |
        WS_VISIBLE
        | BS_PUSHBUTTON, 190, 115, 80, 25, hWnd, (HMENU)CONTROL_START,
        wndEx.hInstance, NULL);
    SendMessage(hStart, WM_SETFONT, (WPARAM)hFont, TRUE);

    HWND hProgress = CreateWindow(PROGRESS_CLASS, _T(""), WS_CHILD
        | WS_VISIBLE | WS_BORDER, 20, 165, 250, 25, hWnd,
        (HMENU)CONTROL_PROGRESS, wndEx.hInstance, NULL);
    SendMessage(hProgress, PBM_SETSTEP, (WPARAM)1, 0);
    SendMessage(hProgress, PBM_SETPOS, (WPARAM)0, 0);

    ShowWindow(hWnd, iWndShow);

    HANDLE hEvent = CreateEvent(NULL, TRUE, FALSE, szEvent);
    HANDLE hMutex = CreateMutex(NULL, FALSE, szMutex);

    MSG msg;

    while (GetMessage(&msg, 0, 0, 0))
    {
        TranslateMessage(&msg);
        DispatchMessage(&msg);
    }

    CloseHandle(hMutex);
    CloseHandle(hEvent);
    UnregisterClass(wndEx.lpszClassName, wndEx.hInstance);

    return (int)msg.wParam;
}
LRESULT CALLBACK WindowProcedure(HWND hWnd, UINT uMsg, WPARAM
    wParam, LPARAM lParam)
{
    switch (uMsg)
    {
        case WM_DESTROY:
        {
            PostQuitMessage(0);
            break;
        }
        case WM_CLOSE:
        {
            HANDLE hEvent = OpenEvent(EVENT_ALL_ACCESS, FALSE, szEvent);
            SetEvent(hEvent);
            CloseHandle(hEvent);

            DestroyWindow(hWnd);
            break;
        }
        case WM_COMMAND:
        {
```

```cpp
                switch (LOWORD(wParam))
                {
                case CONTROL_BROWSE:
                    {
                        if (!FileDialog(hWnd, szFilePath))
                        {
                            MessageBox(hWnd,
                                _T("You must choose valid file path!"),
                                _T("Error!"), MB_OK | MB_TOPMOST | MB_ICONERROR);
                        }
                        else
                        {
                            char szBuffer[MAX_PATH];
                            wsprintfA(szBuffer,
                                "\n File: %s Press \"Start\" now.",
                                strrchr(szFilePath, '\\') + 1);

                            SetWindowTextA(GetDlgItem(hWnd, CONTROL_TEXT),
                                szBuffer);
                        }
                        break;
                    }
                case CONTROL_START:
                    {
                        if (!*szFilePath)
                        {
                            MessageBox(hWnd,
                                _T("You must choose valid file path first!"),
                                _T("Error!"), MB_OK | MB_TOPMOST | MB_ICONERROR);
                            break;
                        }

                        ifstream infile(szFilePath);

                        if (infile.is_open())
                        {
                            string line;
                            while (std::getline(infile, line))
                            {
                                QueueElement* pQElement = new QueueElement();

                                istringstream iss(line);

                                if (!(iss >> pQElement->iA1 >> pQElement->iB1 >>
                                    pQElement->iC1 >> pQElement->iA2 >>
                                    pQElement->iB2 >> pQElement->iC2))
                                {
                                    break;
                                }

                                pQElement->hWndProgress = GetDlgItem(hWnd,
                                    CONTROL_PROGRESS);
                                pQElement->hWndResult = GetDlgItem(hWnd,
                                    CONTROL_RESULT);

                                queue.Enqueue(pQElement);
                            }

                            infile.close();
```

```
                            SendMessage(GetDlgItem(hWnd, CONTROL_PROGRESS),
                                PBM_SETRANGE, 0, MAKELPARAM(0, queue.Count()));

                            SYSTEM_INFO sysInfo;
                            GetSystemInfo(&sysInfo);

                            for (DWORD dwIndex = 0; dwIndex <
                                sysInfo.dwNumberOfProcessors; dwIndex++)
                            {
                                HANDLE hThread = CreateThread(NULL, 0,
                                    (LPTHREAD_START_ROUTINE)StartAddress,
                                    &queue, 0, NULL);
                                SetThreadIdealProcessor(hThread, dwIndex);
                                Sleep(100);
                                CloseHandle(hThread);
                            }
                        }
                        else
                        {
                            MessageBox(hWnd, _T("Cannot open file!"),
                                _T("Error!"), MB_OK | MB_TOPMOST | MB_ICONERROR);
                        }
                        break;
                    }
                }
                break;
            }
            default:
            {
                return DefWindowProc(hWnd, uMsg, wParam, lParam);
            }
        }
        return 0;
    }

    DWORD WINAPI StartAddress(LPVOID lpParameter)
    {
        // 用克莱姆法则解方程组
        CQueue<QueueElement>* pQueue = (CQueue<QueueElement>*)lpParameter;

        HANDLE hEvent = OpenEvent(EVENT_ALL_ACCESS, FALSE, szEvent);
        HANDLE hMutex = OpenMutex(MUTEX_ALL_ACCESS, FALSE, szMutex);
        QueueElement* pQElement = NULL;

        while (true)
        {
            DWORD dwStatus = WaitForSingleObject(hEvent, 10);
            if (dwStatus == WAIT_OBJECT_0)
            {
                break;
            }

            WaitForSingleObject(hMutex, INFINITE);
            pQElement = pQueue->Dequeue();
            ReleaseMutex(hMutex);

            if (pQElement == NULL)
            {
```

```cpp
                break;
            }

            char szBuffer[1024];
            double dDeterminant = (pQElement->iA1 * pQElement->iB2) -
                (pQElement->iB1 * pQElement->iA2);
            if (dDeterminant != 0)
            {
                double dX = ((pQElement->iC1 * pQElement->iB2) -
                    (pQElement->iB1 * pQElement->iC2)) / dDeterminant;
                double dY = ((pQElement->iA1 * pQElement->iC2) -
                    (pQElement->iC1 * pQElement->iA2)) / dDeterminant;
                sprintf_s(szBuffer, " x = %8.4lf,\ty = %8.4lf\n", dX, dY);
            }
            else
            {
                sprintf_s(szBuffer, " Determinant is zero.\n");
            }

            strcat_s(szResult, 4096, szBuffer);
            SetWindowTextA(pQElement->hWndResult, szResult);

            SendMessage(pQElement->hWndProgress, PBM_STEPIT, 0, 0);

            delete pQElement;

            Sleep(1000);
        }

        CloseHandle(hMutex);
        CloseHandle(hEvent);

        return 0L;
    }

BOOL FileDialog(HWND hWnd, LPSTR szFileName)
{
    OPENFILENAMEA ofn;

    char szFile[MAX_PATH];

    ZeroMemory(&ofn, sizeof(ofn));

    ofn.lStructSize = sizeof(ofn);
    ofn.hwndOwner = hWnd;
    ofn.lpstrFile = szFile;
    ofn.lpstrFile[0] = '\0';
    ofn.nMaxFile = sizeof(szFile);
    ofn.lpstrFilter = "All\0*.*\0Text\0*.TXT\0";
    ofn.nFilterIndex = 1;
    ofn.lpstrFileTitle = NULL;
    ofn.nMaxFileTitle = 0;
    ofn.lpstrInitialDir = NULL;
    ofn.Flags = OFN_PATHMUSTEXIST | OFN_FILEMUSTEXIST;

    if (GetOpenFileNameA(&ofn) == TRUE)
    {
        strcpy_s(szFileName, MAX_PATH - 1, szFile);
        return TRUE;
```

```
        }
        return FALSE;
}
```

### 示例分析

和前面的示例类似,我们使用了互斥量对象来同步。但是,本例添加了一个事件来触发已经发生某个动作的线程。这个动作很简单,即用户想结束应用程序。我们给出这个示例,主要是为了让读者了解如何在多线程环境中使用事件对象。

因此,我们在主线程进入消息循环之前创建了一个事件。在 `StartAddress` 线程开始时调用 `WaitForSingleObject`,用 10 毫秒的等待时间,仅为了有时间让线程询问对象的触发状态:如果事件是触发状态,线程必须退出;否则,线程将继续执行。

### 更多讨论

对于多线程环境,事件非常重要。除了同步,那些负责独占访问的对象几乎总是还需要某种触发机制,线程之间将使用这种触发机制进行通信(告知已发生某动作)。我们的示例演示了基本的事件用法,读者可以更深入地使用和研究本书演示的技术。

## 5.8 使用临界区

临界区也是 Windows 的同步对象,其行为与互斥量类似。但是,临界区只能用在单独的进程上下文中,而互斥量可以在多个进程中共享。另外,系统分配临界区对象速度更快,所需的开销更小。

下面的示例中仍使用相同的示例。这次,我们用临界区代替互斥量,能这样做是因为整个应用程序执行在一个进程上下文中。

### 准备就绪

确定安装并运行了 Visual Studio。

### 操作步骤

现在,根据下面的步骤创建程序,稍后再详细解释。

1. 创建一个新的 C++ Win32 空项目,并命名为 `concurrent_operations4`。把第 1 章中的头文件 `CQueue.h` 复制到该项目的目录中。

2. 打开【解决方案资源管理器】,右键单击【头文件】,选择【添加】-【现有项】,添加 `CQueue.h`。

3. 打开【解决方案资源管理器】,右键单击【头文件】,添加新的头文件 main。打开 main.h。

4. 输入下面的代码:

```cpp
#pragma once

#include <windows.h>
#include <commctrl.h>
#include <tchar.h>
#include <psapi.h>
#include <strsafe.h>
#include <string.h>
#include <sstream>
#include <fstream>
#include <commdlg.h>
#include "CQueue.h"

#pragma comment ( lib, "comctl32.lib" )
#pragma comment ( linker, "\"/manifestdependency:type='win32' \
name='Microsoft.Windows.Common-Controls' \
version='6.0.0.0' processorArchitecture='*' \
publicKeyToken='6595b64144ccf1df' language='*'\"" )

using namespace std;

#define CONTROL_BROWSE 100
#define CONTROL_START 101
#define CONTROL_RESULT 103
#define CONTROL_TEXT 104
#define CONTROL_PROGRESS 105

LRESULT CALLBACK WindowProcedure(HWND hWnd, UINT uMsg, WPARAM
    wParam, LPARAM lParam);
DWORD WINAPI StartAddress(LPVOID lpParameter);
BOOL FileDialog(HWND hWnd, LPSTR szFileName);

typedef struct
{
    int iA1;
    int iB1;
    int iC1;
    int iA2;
    int iB2;
    int iC2;
    HWND hWndProgress;
    HWND hWndResult;
} QueueElement, *PQueueElement;
```

5. 打开【解决方案资源管理器】,右键单击【源文件】,添加一个新的源文件 main。打开 main.cpp。

6. 输入下面的代码:

```cpp
#pragma once

/***********************************************************
* 用户指定包含坐标的文本文件 *
* 本例将计算有两个线性方程的方程组 *
```

```
 *      a1x + b1y = c1;                                     *
 *      a2x + b2y = c2;                                     *
 *                                                          *
 *  文本中每行有 6 个坐标                                    *
 *  格式如下:                                               *
 *                                                          *
 *  4 -3 2 -7 5 6                                           *
 *  12 1 -7 9 -5 8                                          *
 *                                                          *
 *  两行数据不要用空行隔开                                  *
 ************************************************************/

#include "main.h"

char szFilePath[MAX_PATH] = { 0 };
char szResult[4096];
CQueue<QueueElement> queue;
TCHAR* szEvent = _T("__event_374__");
CRITICAL_SECTION criticalSection;

int WINAPI _tWinMain(HINSTANCE hThis, HINSTANCE hPrev, LPTSTR
    szCommandLine, int iWndShow)
{
    UNREFERENCED_PARAMETER(hPrev);
    UNREFERENCED_PARAMETER(szCommandLine);

    TCHAR* szWindowClass = _T("__concurrent_operations__");
    WNDCLASSEX wndEx = { 0 };

    wndEx.cbSize = sizeof(WNDCLASSEX);
    wndEx.style = CS_HREDRAW | CS_VREDRAW;
    wndEx.lpfnWndProc = WindowProcedure;
    wndEx.cbClsExtra = 0;
    wndEx.cbWndExtra = 0;
    wndEx.hInstance = hThis;
    wndEx.hIcon = LoadIcon(wndEx.hInstance, MAKEINTRESOURCE(IDI_APPLICATION));
    wndEx.hCursor = LoadCursor(NULL, IDC_ARROW);
    wndEx.hbrBackground = (HBRUSH)(COLOR_WINDOW + 1);
    wndEx.lpszMenuName = NULL;
    wndEx.lpszClassName = szWindowClass;
    wndEx.hIconSm = LoadIcon(wndEx.hInstance, MAKEINTRESOURCE(IDI_APPLICATION));

    if (!RegisterClassEx(&wndEx))
    {
        return 1;
    }

    InitCommonControls();

    HWND hWnd = CreateWindow(szWindowClass,
        _T("Concurrent operations"), WS_OVERLAPPED
        | WS_CAPTION | WS_SYSMENU, 50, 50, 305, 250, NULL, NULL,
        wndEx.hInstance, NULL);

    if (!hWnd)
    {
        return NULL;
    }
```

```
HFONT hFont = CreateFont(14, 0, 0, 0, FW_NORMAL, FALSE,
    FALSE, FALSE, BALTIC_CHARSET, OUT_DEFAULT_PRECIS,
    CLIP_DEFAULT_PRECIS, DEFAULT_QUALITY, DEFAULT_PITCH
    | FF_MODERN, _T("Microsoft Sans Serif"));

HWND hLabel = CreateWindow(_T("STATIC"),
    _T(" Press \"Browse\" and choose file with coordinates: "),
    WS_CHILD | WS_VISIBLE | SS_CENTERIMAGE | SS_LEFT | WS_BORDER,
    20, 20, 250, 25, hWnd, (HMENU)CONTROL_TEXT, wndEx.hInstance,
    NULL);
SendMessage(hLabel, WM_SETFONT, (WPARAM)hFont, TRUE);

HWND hResult = CreateWindow(_T("STATIC"), _T(""), WS_CHILD |
    WS_VISIBLE | SS_LEFT | WS_BORDER, 20, 65, 150, 75, hWnd,
    (HMENU)CONTROL_RESULT, wndEx.hInstance, NULL);
SendMessage(hResult, WM_SETFONT, (WPARAM)hFont, TRUE);

HWND hBrowse = CreateWindow(_T("BUTTON"), _T("Browse"), WS_CHILD
    | WS_VISIBLE
    | BS_PUSHBUTTON, 190, 65, 80, 25, hWnd, (HMENU)CONTROL_BROWSE,
    wndEx.hInstance, NULL);
SendMessage(hBrowse, WM_SETFONT, (WPARAM)hFont, TRUE);

HWND hStart = CreateWindow(_T("BUTTON"), _T("Start"), WS_CHILD |
    WS_VISIBLE
    | BS_PUSHBUTTON, 190, 115, 80, 25, hWnd, (HMENU)CONTROL_START,
    wndEx.hInstance, NULL);
SendMessage(hStart, WM_SETFONT, (WPARAM)hFont, TRUE);

HWND hProgress = CreateWindow(PROGRESS_CLASS, _T(""), WS_CHILD |
    WS_VISIBLE | WS_BORDER,
    20, 165, 250, 25, hWnd, (HMENU)CONTROL_PROGRESS, wndEx.
    hInstance, NULL);
SendMessage(hProgress, PBM_SETSTEP, (WPARAM)1, 0);
SendMessage(hProgress, PBM_SETPOS, (WPARAM)0, 0);

ShowWindow(hWnd, iWndShow);

HANDLE hEvent = CreateEvent(NULL, TRUE, FALSE, szEvent);
InitializeCriticalSection(&criticalSection);

MSG msg;
while (GetMessage(&msg, 0, 0, 0))
{
    TranslateMessage(&msg);
    DispatchMessage(&msg);
}

DeleteCriticalSection(&criticalSection);
CloseHandle(hEvent);
UnregisterClass(wndEx.lpszClassName, wndEx.hInstance);

return (int)msg.wParam;
}

LRESULT CALLBACK WindowProcedure(HWND hWnd, UINT uMsg, WPARAM
    wParam, LPARAM lParam)
{
    switch (uMsg)
```

```cpp
{
    case WM_DESTROY:
    {
        PostQuitMessage(0);
        break;
    }
    case WM_CLOSE:
    {
        HANDLE hEvent = OpenEvent(EVENT_ALL_ACCESS, FALSE, szEvent);
        SetEvent(hEvent);
        CloseHandle(hEvent);

        DestroyWindow(hWnd);
        break;
    }
    case WM_COMMAND:
    {
        switch (LOWORD(wParam))
        {
            case CONTROL_BROWSE:
            {
                if (!FileDialog(hWnd, szFilePath))
                {
                    MessageBox(hWnd,
                        _T("You must choose valid file path!"),
                        _T("Error!"), MB_OK | MB_TOPMOST | MB_ICONERROR);
                }
                else
                {
                    char szBuffer[MAX_PATH];
                    wsprintfA(szBuffer,
                        "\n File: %s Press \"Start\" now.",
                        strrchr(szFilePath, '\\') + 1);
                    SetWindowTextA(GetDlgItem(hWnd, CONTROL_TEXT),
                        szBuffer);
                }
                break;
            }
            case CONTROL_START:
            {
                if (!*szFilePath)
                {
                    MessageBox(hWnd,
                        _T("You must choose valid file path first!"),
                        _T("Error!"), MB_OK | MB_TOPMOST | MB_ICONERROR);
                    break;
                }

                ifstream infile(szFilePath);

                if (infile.is_open())
                {
                    string line;
                    while (std::getline(infile, line))
                    {
                        QueueElement* pQElement = new QueueElement();
                        istringstream iss(line);
                        if (!(iss >> pQElement->iA1 >> pQElement->iB1 >>
                            pQElement->iC1 >> pQElement->iA2 >>
```

```
                            pQElement->iB2 >> pQElement->iC2))
                    {
                        break;
                    }

                    pQElement->hWndProgress = GetDlgItem(hWnd,
                        CONTROL_PROGRESS);
                    pQElement->hWndResult = GetDlgItem(hWnd,
                        CONTROL_RESULT);
                    queue.Enqueue(pQElement);
                }

                infile.close();

                SendMessage(GetDlgItem(hWnd, CONTROL_PROGRESS),
                    PBM_SETRANGE, 0, MAKELPARAM(0, queue.Count()));

                SYSTEM_INFO sysInfo;
                GetSystemInfo(&sysInfo);

                for (DWORD dwIndex = 0; dwIndex <
                    sysInfo.dwNumberOfProcessors; dwIndex++)
                {
                    HANDLE hThread = CreateThread(NULL, 0,
                        (LPTHREAD_START_ROUTINE)StartAddress,
                        &queue, 0, NULL);
                    SetThreadIdealProcessor(hThread, dwIndex);

                    Sleep(100);
                    CloseHandle(hThread);
                }
            }
            else
            {
                MessageBox(hWnd, _T("Cannot open file!"),
                    _T("Error!"), MB_OK | MB_TOPMOST | MB_ICONERROR);
            }
            break;
        }
    }
    break;
    default:
    {
        return DefWindowProc(hWnd, uMsg, wParam, lParam);
    }
    }
    return 0;
}

DWORD WINAPI StartAddress(LPVOID lpParameter)
{
    // 用克莱姆法则解方程组
    CQueue<QueueElement>* pQueue = (CQueue<QueueElement>*)lpParameter;

    HANDLE hEvent = OpenEvent(EVENT_ALL_ACCESS, FALSE, szEvent);
    QueueElement* pQElement = NULL;

    while (true)
```

```
        {
            DWORD dwStatus = WaitForSingleObject(hEvent, 10);

            if (dwStatus == WAIT_OBJECT_0)
            {
                break;
            }

            EnterCriticalSection(&criticalSection);

            pQElement = pQueue->Dequeue();

            LeaveCriticalSection(&criticalSection);

            if (pQElement == NULL)
            {
                break;
            }

            char szBuffer[1024];

            double dDeterminant = (pQElement->iA1 * pQElement->iB2) -
                (pQElement->iB1 * pQElement->iA2);

            if (dDeterminant != 0)
            {
                double dX = ((pQElement->iC1 * pQElement->iB2) -
                    (pQElement->iB1 * pQElement->iC2)) / dDeterminant;
                double dY = ((pQElement->iA1 * pQElement->iC2) -
                    (pQElement->iC1 * pQElement->iA2)) / dDeterminant;

                sprintf_s(szBuffer, " x = %8.4lf,\ty = %8.4lf\n", dX, dY);
            }
            else
            {
                sprintf_s(szBuffer, " Determinant is zero.\n");
            }

            strcat_s(szResult, 4096, szBuffer);
            SetWindowTextA(pQElement->hWndResult, szResult);

            SendMessage(pQElement->hWndProgress, PBM_STEPIT, 0, 0);
            delete pQElement;

            Sleep(1000);
        }

        CloseHandle(hEvent);

        return 0L;
}

BOOL FileDialog(HWND hWnd, LPSTR szFileName)
{
    OPENFILENAMEA ofn;

    char szFile[MAX_PATH];

    ZeroMemory(&ofn, sizeof(ofn));
```

```
        ofn.lStructSize = sizeof(ofn);
        ofn.hwndOwner = hWnd;
        ofn.lpstrFile = szFile;
        ofn.lpstrFile[0] = '\0';
        ofn.nMaxFile = sizeof(szFile);
        ofn.lpstrFilter = "All\0*.*\0Text\0*.TXT\0";
        ofn.nFilterIndex = 1;
        ofn.lpstrFileTitle = NULL;
        ofn.nMaxFileTitle = 0;
        ofn.lpstrInitialDir = NULL;
        ofn.Flags = OFN_PATHMUSTEXIST | OFN_FILEMUSTEXIST;

        if (GetOpenFileNameA(&ofn) == TRUE)
        {
            strcpy_s(szFileName, MAX_PATH - 1, szFile);
            return TRUE;
        }

        return FALSE;
    }
```

### 示例分析

本例与前面示例的区别在于，使用了不同的同步对象。对我们而言重要的是，临界区对象只允许一个线程执行独占访问，这与互斥量一样，但是临界区的效率更高。

### 更多讨论

选互斥量还是临界区要特别谨慎。除了要考虑是否需要在进程间共享同步对象、在提供正确行为的前提下系统的运行速度，以及创建和操控的开销外，在选择合适的同步对象和技术前，还要尽量预测程序在使用中出现的各种情况。

## 5.9 使用管道

讲到同步，绝对有必要介绍一个多进程示例。与在单独的地址空间中运行多线程不同，每个进程都有不同的优点和缺点。

通常，多进程的优点是出错的概率低。进程的上下文和地址属于进程以及所有资源私有（包括线程）。在一个进程中分配一些内存空间，并将其指针传递给另一个进程，是毫无意义的。操作系统为每个进程都创建了单独的地址空间，正在运行的进程并不知道其他进程的任何情况。而且进程很贪婪，它想获得自己所需的所有资源。因此，在运行多进程时，需要一块共享内存区域来共享数据。或者说，需要管道进行通信。另外，还有一种通信对象叫做套接字（socket），但是这部分内容超出了本书讨论的范围。本节只介绍管道，因为它比其他方法更快，更便于使用。

Windows套接字让程序员能创建高级的因特网、内部网和其他带网络功能的应用程序，通过网线传输应用程序数据，不依赖正在使用的网络协议。

多进程的缺点是有许多更麻烦的事情，因为必须把要在进程间共享的所有内容都写入共享内存，这样其他进程才能读取。我们仍沿用相同的例子，但是这次使用工作进程。这种方案在守护任务（*daemon task*）中很有用，如服务器或打印机。

## 准备就绪

确定安装并运行了 Visual Studio。

## 操作步骤

现在，根据下面的步骤创建程序，稍后再详细解释。

1. 创建一个新的 C++ Win32 空项目，并命名为 `concurrent_operations5`。把第 1 章中的头文件 `CQueue.h` 复制到该项目的目录中。

2. 打开【解决方案资源管理器】，右键单击【头文件】，选择【添加】-【现有项】，添加 `CQueue.h`。

3. 打开【解决方案资源管理器】，右键单击【头文件】，添加新的头文件 `main`。打开 `main.h`。

4. 输入下面的代码：

```
#pragma once

#include <windows.h>
#include <commctrl.h>
#include <tchar.h>
#include <psapi.h>
#include <strsafe.h>
#include <string.h>
#include <sstream>
#include <fstream>
#include <commdlg.h>
#include "CQueue.h"

#pragma comment ( linker, "\"/manifestdependency:type='win32' \
name='Microsoft.Windows.Common-Controls' \
version='6.0.0.0' processorArchitecture='*' \
publicKeyToken='6595b64144ccf1df' language='*'\"" )

using namespace std;

#define CONTROL_BROWSE          100
```

```
#define CONTROL_START        101
#define CONTROL_RESULT       103
#define CONTROL_TEXT         104
#define CONTROL_PROGRESS     105

#define BUFFER_SIZE          1024

#define PIPE_NAME    _T( "\\\\.\\pipe\\__pipe_636__" )
#define EVENT_NAME   _T( "__event_879__" )
#define MUTEX_NAME   _T( "__mutex_132__" )
#define MAPPING_NAME _T( "__mapping_514__" )

typedef struct
{
    int iA1;
    int iB1;
    int iC1;
    int iA2;
    int iB2;
    int iC2;
    HWND hWndProgress;
    HWND hWndResult;
} QueueElement, *PQueueElement;

LRESULT CALLBACK WindowProcedure(HWND hWnd, UINT uMsg, WPARAM
    wParam, LPARAM lParam);
DWORD WINAPI ListenerRoutine(LPVOID lpParameter);
DWORD WINAPI StartAddress(LPVOID lpParameter);
BOOL FileDialog(HWND hWnd, LPSTR szFileName);
bool CalculateCramer(QueueElement* pQElement, char* szResult);
BOOL StartProcess(HWND hWndResult);
```

5. 打开【解决方案资源管理器】，右键单击【源文件】，添加一个新的源文件 main。打开 main.cpp。

6. 输入下面的代码：

```
#pragma once

/***********************************************************
* 用户指定包含坐标的文本文件 *
* 本例将计算有两个线性方程的方程组 *
* a1x + b1y = c1; *
* a2x + b2y = c2; *
* *
* 文本中每行有6个坐标 *
* 格式如下： *
* *
* 4 -3 2 -7 5 6 *
* 12 1 -7 9 -5 8 *
* * *
* 两行数据不要用空行隔开 *
***********************************************************/

#include "main.h"

char szFilePath[MAX_PATH] = { 0 };
char szResult[4096];
```

```
CQueue<QueueElement> queue;

int WINAPI _tWinMain(HINSTANCE hThis, HINSTANCE hPrev, LPTSTR
    szCommandLine, int iWndShow)
{
    UNREFERENCED_PARAMETER(hPrev);
    UNREFERENCED_PARAMETER(szCommandLine);

    TCHAR* szWindowClass = _T("__concurrent_operations__");
    WNDCLASSEX wndEx = { 0 };

    wndEx.cbSize = sizeof(WNDCLASSEX);
    wndEx.style = CS_HREDRAW | CS_VREDRAW;
    wndEx.lpfnWndProc = WindowProcedure;
    wndEx.cbClsExtra = 0;
    wndEx.cbWndExtra = 0;
    wndEx.hInstance = hThis;
    wndEx.hIcon = LoadIcon(wndEx.hInstance, MAKEINTRESOURCE(IDI_APPLICATION));
    wndEx.hCursor = LoadCursor(NULL, IDC_ARROW);
    wndEx.hbrBackground = (HBRUSH)(COLOR_WINDOW + 1);
    wndEx.lpszMenuName = NULL;
    wndEx.lpszClassName = szWindowClass;
    wndEx.hIconSm = LoadIcon(wndEx.hInstance, MAKEINTRESOURCE(IDI_APPLICATION));

    if (!RegisterClassEx(&wndEx))
    {
        return 1;
    }

    InitCommonControls();

    HWND hWnd = CreateWindow(szWindowClass,
        _T("Concurrent operations"), WS_OVERLAPPED
        | WS_CAPTION | WS_SYSMENU, 50, 50, 305, 250, NULL, NULL,
        wndEx.hInstance, NULL);

    if (!hWnd)
    {
        return NULL;
    }

    HFONT hFont = CreateFont(14, 0, 0, 0, FW_NORMAL, FALSE,
        FALSE, FALSE, BALTIC_CHARSET, OUT_DEFAULT_PRECIS,
        CLIP_DEFAULT_PRECIS, DEFAULT_QUALITY, DEFAULT_PITCH
        | FF_MODERN, _T("Microsoft Sans Serif"));

    HWND hLabel = CreateWindow(_T("STATIC"),
        _T(" Press \"Browse\" and choose file with coordinates: "),
        WS_CHILD | WS_VISIBLE | SS_CENTERIMAGE | SS_LEFT | WS_BORDER,
        20, 20, 250, 25, hWnd, (HMENU)CONTROL_TEXT, wndEx.hInstance, NULL);

    SendMessage(hLabel, WM_SETFONT, (WPARAM)hFont, TRUE);

    HWND hResult = CreateWindow(_T("STATIC"), _T(""), WS_CHILD |
        WS_VISIBLE | SS_LEFT | WS_BORDER, 20, 65, 150, 75, hWnd,
        (HMENU)CONTROL_RESULT, wndEx.hInstance, NULL);
    SendMessage(hResult, WM_SETFONT, (WPARAM)hFont, TRUE);

    HWND hBrowse = CreateWindow(_T("BUTTON"), _T("Browse"), WS_CHILD
```

```
                | WS_VISIBLE
                | BS_PUSHBUTTON, 190, 65, 80, 25, hWnd, (HMENU)CONTROL_BROWSE,
                wndEx.hInstance, NULL);
            SendMessage(hBrowse, WM_SETFONT, (WPARAM)hFont, TRUE);

            HWND hStart = CreateWindow(_T("BUTTON"), _T("Start"), WS_CHILD |
                WS_VISIBLE
                | BS_PUSHBUTTON, 190, 115, 80, 25, hWnd, (HMENU)CONTROL_START,
                wndEx.hInstance, NULL);
            SendMessage(hStart, WM_SETFONT, (WPARAM)hFont, TRUE);

            HWND hProgress = CreateWindow(PROGRESS_CLASS, _T(""), WS_CHILD
                | WS_VISIBLE | WS_BORDER, 20, 165, 250, 25, hWnd,
                (HMENU)CONTROL_PROGRESS, wndEx.hInstance, NULL);
            SendMessage(hProgress, PBM_SETSTEP, (WPARAM)1, 0);
            SendMessage(hProgress, PBM_SETPOS, (WPARAM)0, 0);

            void* params[2] = { hProgress, hResult };
            HANDLE hThread = CreateThread(NULL, 0, (LPTHREAD_START_ROUTINE)
                ListenerRoutine, params, 0, NULL);
            Sleep(100);
            CloseHandle(hThread);

            ShowWindow(hWnd, iWndShow);

            HANDLE hEvent = CreateEvent(NULL, TRUE, FALSE, EVENT_NAME);
            HANDLE hMutex = CreateMutex(NULL, FALSE, MUTEX_NAME);
            HANDLE hMaping = CreateFileMapping((HANDLE)-1, NULL,
                PAGE_READWRITE, 0, sizeof(QueueElement), MAPPING_NAME);

            MSG msg;
            while (GetMessage(&msg, 0, 0, 0))
            {
                TranslateMessage(&msg);
                DispatchMessage(&msg);
            }

            CloseHandle(hMaping);
            CloseHandle(hMutex);
            CloseHandle(hEvent);
            UnregisterClass(wndEx.lpszClassName, wndEx.hInstance);

            return (int)msg.wParam;
        }
        LRESULT CALLBACK WindowProcedure(HWND hWnd, UINT uMsg, WPARAM
            wParam, LPARAM lParam)
        {
            switch (uMsg)
            {
                case WM_DESTROY:
                {
                    PostQuitMessage(0);
                    break;
                }
                case WM_CLOSE:
                {
                    HANDLE hEvent = OpenEvent(EVENT_ALL_ACCESS, FALSE,
                        EVENT_NAME);
```

```cpp
                SetEvent(hEvent);
                CloseHandle(hEvent);

                DestroyWindow(hWnd);
                break;
            }
            case WM_COMMAND:
            {
                switch (LOWORD(wParam))
                {
                    case CONTROL_BROWSE:
                    {
                        if (!FileDialog(hWnd, szFilePath))
                        {
                            MessageBox(hWnd,
                                _T("You must choose valid file path!"),
                                _T("Error!"), MB_OK | MB_TOPMOST | MB_ICONERROR);
                        }
                        else
                        {
                            char szBuffer[MAX_PATH];
                            wsprintfA(szBuffer,
                                "\n File: %s Press \"Start\" now.",
                                strrchr(szFilePath, '\\') + 1);

                            SetWindowTextA(GetDlgItem(hWnd, CONTROL_TEXT),
                                szBuffer);
                        }
                        break;
                    }
                    case CONTROL_START:
                    {
                        if (!*szFilePath)
                        {
                            MessageBox(hWnd,
                                _T("You must choose valid file path first!"),
                                _T("Error!"), MB_OK | MB_TOPMOST | MB_ICONERROR);
                            break;
                        }
                        ifstream infile(szFilePath);
                        if (infile.is_open())
                        {
                            string line;
                            while (std::getline(infile, line))
                            {
                                QueueElement* pQElement = new QueueElement();
                                istringstream iss(line);

                                if (!(iss >> pQElement->iA1 >> pQElement->iB1 >>
                                    pQElement->iC1 >> pQElement->iA2 >>
                                    pQElement->iB2 >> pQElement->iC2))
                                {
                                    break;
                                }

                                pQElement->hWndProgress = GetDlgItem(hWnd,
                                    CONTROL_PROGRESS);
                                pQElement->hWndResult = GetDlgItem(hWnd,
                                    CONTROL_RESULT);
```

```cpp
                                queue.Enqueue(pQElement);
                            }
                            infile.close();
                            SendMessage(GetDlgItem(hWnd, CONTROL_PROGRESS),
                                PBM_SETRANGE, 0, MAKELPARAM(0, queue.Count()));

                            int iCount = queue.Count();
                            for (int iIndex = 0; iIndex < iCount; iIndex++)
                            {
                                StartProcess(GetDlgItem(hWnd, CONTROL_TEXT));
                                Sleep(100);
                            }
                        }
                        else
                        {
                            MessageBox(hWnd, _T("Cannot open file!"),
                                _T("Error!"), MB_OK | MB_TOPMOST | MB_ICONERROR);
                        }
                        break;
                    }
                }
                break;
            }
        default:
            {
                return DefWindowProc(hWnd, uMsg, wParam, lParam);
            }
        }
        return 0;
    }

    DWORD WINAPI ListenerRoutine(LPVOID lpParameter)
    {
        void** lpParameters = (void**)lpParameter;
        HWND hWndProgress = (HWND)lpParameters[0];
        HWND hWndResult = (HWND)lpParameters[1];

        HANDLE hEvent = OpenEvent(EVENT_ALL_ACCESS, FALSE, EVENT_NAME);

        while (TRUE)
        {
            DWORD dwStatus = WaitForSingleObject(hEvent, INFINITE);
            if (dwStatus == WAIT_OBJECT_0)
            {
                break;
            }

            HANDLE hPipe = CreateNamedPipe(PIPE_NAME, PIPE_ACCESS_DUPLEX,
                PIPE_WAIT | PIPE_READMODE_MESSAGE
                | PIPE_TYPE_MESSAGE, PIPE_UNLIMITED_INSTANCES, BUFFER_SIZE,
                BUFFER_SIZE, 0, NULL);

            if (hPipe == INVALID_HANDLE_VALUE)
            {
                char szBuffer[MAX_PATH];
                wsprintfA(szBuffer, " Error: [%u]\n", GetLastError());
                SetWindowTextA(hWndResult, szBuffer);
```

```
                return 2L;
            }
            if (ConnectNamedPipe(hPipe, NULL))
            {
                void* params[2] = { hPipe, hWndResult };
                HANDLE hThread = CreateThread(NULL, 0,
                    (LPTHREAD_START_ROUTINE)StartAddress, params, 0, NULL);
                Sleep(100);
                CloseHandle(hThread);
            }

            SendMessage(hWndProgress, PBM_STEPIT, 0, 0);
        }

    CloseHandle(hEvent);
    return 0L;
}

DWORD WINAPI StartAddress(LPVOID lpParameter)
{
    void** lpParameters = (void**)lpParameter;
    HANDLE hPipe = (HANDLE)lpParameters[0];
    HWND hWndResult = (HWND)lpParameters[1];

    char* szRequest = (char*)HeapAlloc(GetProcessHeap(), 0,
        BUFFER_SIZE * sizeof(char));
    char* szReply = (char*)HeapAlloc(GetProcessHeap(), 0,
        BUFFER_SIZE * sizeof(char));

    DWORD dwBytesRead = 0;
    DWORD dwReplyBytes = 0;
    DWORD dwBytesWritten = 0;

    memset(szRequest, 0, BUFFER_SIZE * sizeof(char));
    memset(szReply, 0, BUFFER_SIZE * sizeof(char));

    if (!ReadFile(hPipe, szRequest, BUFFER_SIZE * sizeof(char),
        &dwBytesRead, NULL))
    {
        return 2L;
    }

    char szBuffer[1024] = { 0 };
    wsprintfA(szBuffer, " PID: [%s] connected.\n", szRequest);
    strcat_s(szResult, szBuffer);
    SetWindowTextA(hWndResult, szResult);

    HANDLE hMutex = OpenMutex(MUTEX_ALL_ACCESS, FALSE, MUTEX_NAME);

    WaitForSingleObject(hMutex, INFINITE);

    QueueElement* pQElement = (QueueElement*)queue.Dequeue();

    HANDLE hMapping = OpenFileMapping(FILE_MAP_ALL_ACCESS, FALSE,
        MAPPING_NAME);
    QueueElement* pMapping = (QueueElement*)MapViewOfFile(hMapping,
        FILE_MAP_ALL_ACCESS, 0, 0, 0);

    memcpy(pMapping, pQElement, sizeof(QueueElement));
```

```
            UnmapViewOfFile(pMapping);
            CloseHandle(hMapping);

            ReleaseMutex(hMutex);
            CloseHandle(hMutex);

            sprintf_s(szReply, BUFFER_SIZE * sizeof(char), "OK%s",
                szRequest);
            dwReplyBytes = (DWORD)((strlen(szReply) + 1) * sizeof(char));

            if (!WriteFile(hPipe, szReply, dwReplyBytes, &dwBytesWritten,
                NULL))
            {
                return 2L;
            }

            memset(szRequest, 0, BUFFER_SIZE * sizeof(char));

            if (!ReadFile(hPipe, szRequest, BUFFER_SIZE * sizeof(char),
                &dwBytesRead, NULL))
            {
                return 2L;
            }

            wsprintfA(szBuffer, "%s", szRequest);
            strcat_s(szResult, szBuffer);

            SetWindowTextA(hWndResult, szResult);

            delete pQElement;

            FlushFileBuffers(hPipe);
            DisconnectNamedPipe(hPipe);
            CloseHandle(hPipe);
            HeapFree(GetProcessHeap(), 0, szRequest);
            HeapFree(GetProcessHeap(), 0, szReply);

            return 0L;
    }

    BOOL FileDialog(HWND hWnd, LPSTR szFileName)
    {
        OPENFILENAMEA ofn;

        char szFile[MAX_PATH];

        ZeroMemory(&ofn, sizeof(ofn));

        ofn.lStructSize = sizeof(ofn);
        ofn.hwndOwner = hWnd;
        ofn.lpstrFile = szFile;
        ofn.lpstrFile[0] = '\0';
        ofn.nMaxFile = sizeof(szFile);
        ofn.lpstrFilter = "All\0*.*\0Text\0*.TXT\0";
        ofn.nFilterIndex = 1;
        ofn.lpstrFileTitle = NULL;
        ofn.nMaxFileTitle = 0;
        ofn.lpstrInitialDir = NULL;
```

```cpp
        ofn.Flags = OFN_PATHMUSTEXIST | OFN_FILEMUSTEXIST;

        if (GetOpenFileNameA(&ofn) == TRUE)
        {
            strcpy_s(szFileName, MAX_PATH - 1, szFile);
            return TRUE;
        }

        return FALSE;
    }

    BOOL StartProcess(HWND hWndResult)
    {
        STARTUPINFO startupInfo = { 0 };
        PROCESS_INFORMATION processInformation = { 0 };

        BOOL bSuccess = CreateProcess(
            _T("..\\x64\\Debug\\ClientProcess.exe"), NULL, NULL, NULL,
            FALSE, 0, NULL, NULL, &startupInfo, &processInformation);

        if (!bSuccess)
        {
            char szBuffer[MAX_PATH];
            wsprintfA(szBuffer, " Process creation fails: [%u]\n",
                GetLastError());
            SetWindowTextA(hWndResult, szBuffer);
        }

        return bSuccess;
    }
```

7. 打开【解决方案资源管理器】，右键单击【解决方案 "concurrent_operations5"】，添加一个新的空控制台应用程序 ClientProcess。然后，右键单击 ClientProcess 项目，添加一个新的源文件 main。打开 main.cpp。

8. 输入下面的代码：

```cpp
#pragma once

#include "..\concurrent_operations5\main.h"

int _tmain(void)
{
    char szBuffer[BUFFER_SIZE];
    DWORD dwRead = 0;
    DWORD dwWritten = 0;
    DWORD dwMode = PIPE_READMODE_MESSAGE;

    char szMessage[1024] = { 0 };
    wsprintfA(szMessage, "%u", GetCurrentProcessId());

    DWORD dwToWrite = (DWORD)(strlen(szMessage) * sizeof(char));

    HANDLE hPipe = NULL;

    while (true)
    {
```

```c
            hPipe = CreateFile(PIPE_NAME, GENERIC_READ | GENERIC_WRITE, 0,
                NULL, OPEN_EXISTING, 0, NULL);
            if (hPipe != INVALID_HANDLE_VALUE)
            {
                WaitNamedPipe(PIPE_NAME, INFINITE);
                SetNamedPipeHandleState(hPipe, &dwMode, NULL, NULL);
                break;
            }

            Sleep(100);
        }

        if (!WriteFile(hPipe, szMessage, dwToWrite, &dwWritten, NULL))
        {
            printf("Error: [%u]\n", GetLastError());
            return system("pause");
        }

        printf("Request sent.\nReceiving task:\n");
        char szResult[1024] = { 0 };

        while (ReadFile(hPipe, szBuffer, BUFFER_SIZE * sizeof(char),
            &dwRead, NULL) && GetLastError() != ERROR_MORE_DATA)
        {
            if (szBuffer[0] == 'O' && szBuffer[1] == 'K' && ((DWORD)
                strtol(szBuffer + 2, NULL, 10) == GetCurrentProcessId()))
            {
                printf("Client process: [%s]\n", szBuffer + 2);

                HANDLE hMutex = OpenMutex(MUTEX_ALL_ACCESS,
                    FALSE, MUTEX_NAME);
                WaitForSingleObject(hMutex, INFINITE);

                HANDLE hMapping = OpenFileMapping(FILE_MAP_ALL_ACCESS,
                    FALSE, MAPPING_NAME);
                QueueElement* pMapping = (QueueElement*)
                    MapViewOfFile(hMapping, FILE_MAP_ALL_ACCESS, 0, 0, 0);

                if (pMapping != 0)
                {
                    if (CalculateCramer(pMapping, szResult))
                    {
                        dwToWrite = (DWORD)(strlen(szResult) * sizeof(char));

                        if (!WriteFile(hPipe, szResult, dwToWrite, &dwWritten,
                            NULL))
                        {
                            printf("Error: [%u]\n", GetLastError());
                            break;
                        }
                        else
                        {
                            printf("Result: %s\n", szResult);
                        }
                    }
                }

                UnmapViewOfFile(pMapping);
                CloseHandle(hMapping);
```

```cpp
                ReleaseMutex(hMutex);
                CloseHandle(hMutex);
            }
            else
            {
                printf("Error in connection [%u]\n", GetLastError());
            }
        }

        printf("\nSuccess!\nPress ENTER to exit.");
        scanf_s("%c", szBuffer);

        CloseHandle(hPipe);
        return 0;
    }

    bool CalculateCramer(QueueElement* pQElement, char* szResult)
    {
        // 用克莱姆法则解线性方程组
        double dDeterminant = (pQElement->iA1 * pQElement->iB2) -
            (pQElement->iB1 * pQElement->iA2);

        if (dDeterminant != 0)
        {
            double dX = ((pQElement->iC1 * pQElement->iB2) -
                (pQElement->iB1 * pQElement->iC2)) / dDeterminant;
            double dY = ((pQElement->iA1 * pQElement->iC2) -
                (pQElement->iC1 * pQElement->iA2)) / dDeterminant;
            sprintf_s(szResult, strlen(szResult) - 1,
                " x = %8.4lf,\ty = %8.4lf\n", dX, dY);
        }
        else
        {
            sprintf_s(szResult, strlen(szResult) - 1,
                " Determinant is zero.\n");
        }

        return true;
    }
```

### 示例分析

本例的代码较多，不过多出来的代码才是关键。本例没有任何线程间的安全措施（即没有进行同步），某个线程会把其他线程的数据或全局变量弄乱。但是，使用多进程会更安全。我们的意思是，可以根据进程自身的地址空间、全局变量和同步对象来划分进程。另一方面，如果一个线程被阻塞，等待它输出或共享资源的所有线程都会被阻塞。学到同步的时候，理解进程模型和进程间通信（IPC）也很重要。接下来，我们先来看 main.h。由于很多项目都要使用管道名、互斥量名和文件映射名，所以我们在 main 头文件中把它们定义为宏。其他内容和原来的 main 头文件相同。

现在，分析 concurrent_operations5。该应用程序的入口点和原来相同。我们在主线程进入消息循环之前，创建了事件、互斥量和文件映射；在应用程序关闭后妥善处理了所有的相关对象。关于文件映射，还有一件重要的事情是正确设置共享对象的大小，我们用 sizeof(QueueElement) 来设置。

我们还创建了一个**监听线程**（*listener thread*），用 ListenerRoutine 作为开始地址。它的任务是无限循环，或者在事件变成 signaled 之前一直循环。它将创建一个已命名的管道，并等待客户端通过调用 ConnectNamedPipe API 连接（第 4 章介绍过）。

当客户端连接时，它将启动另一个线程处理客户的请求，并等待下一个连接。工作线程从 StartAddress 例程开始，它的任务是为请求和回复分配足够的空间，然后调用 ReadFile API，在客户写入请求并将请求返回之前一直等待。再次强调，重点是要确定保护什么内容。如果多进程要读取和写入共享文件映射，那就是必须保护的地方！

需要执行的步骤是，获得互斥量的所有权，从队列获取元素，写入共享文件映射，然后释放互斥量。当回复准备好时，将使用 WriteFile API 写入管道，并再次等待客户的回复（使用 ReadFile）。接收到外部进程的计算结果后，释放已用的资源并退出。

我们修改了 WindowProcedure 中的内容，创建了与队列中元素数量相同的进程，而不是线程。或许只用一个工作进程负责计算，然后逐个接收任务更好。但是，为了突出进程间通信和同步的重要部分，我们才故意这样设计。

接下来，我们分析客户进程。首先，它尝试连接管道。读者可能认为主应用程序创建进程很快（只需 0.1 秒），这些进程很可能同时连接管道。其实，这是不可能的。即使出现这样的情况，你也会收到 ERROR_INVALID_HANDLE 错误。这就是进程需要循环的原因，在被连接之前延迟 0.1 秒。连接成功后，进程将使用唯一的进程 ID 标识自己。然后，使用 ReadFile API，在请求到达之前一直等待。如果请求格式是 OK12345（这里，12345 表示它的进程标识符），它将询问互斥量的所有权。当进程获得所有权时，它将从共享文件映射中读取坐标，并使用 WriteFile 再次响应。响应成功后，它将释放互斥量并把结果写入控制台输出中。

## 更多讨论

本章介绍的这些技术算是抛砖引玉，读者今后会用到这些知识。希望读者通过本章的学习更好地理解设计和开发过程中的一些重要步骤，并以此作为将来开发的起点。如果开发复杂的系统（任务），还必须深入学习和研究操作系统。再次强调，一个好的设计对于应用程序来说至关重要！记住在设计过程中要格外仔细，特别是在需要并行执行的地方。

# 第 6 章 .NET 框架中的线程

**本章介绍以下内容：**
- 托管代码和非托管代码
- 如何在.NET 中运行线程
- 前台线程和后台线程的区别
- 理解.NET 同步要素
- 锁和避免死锁
- 线程安全和.NET 框架的类型
- 用事件等待句柄触发
- 基于事件的异步模式
- BackgroudWorker 类
- 中断、中止和安全取消线程执行
- 非阻塞同步
- 用 Wait 和 Pulse 发信号
- Barrier 类

## 6.1 介绍

现在，市面上有许多供程序员开发使用的框架。这些框架提供了许多特性和工具，改善了用户体验，让编程更加轻松。

本章将介绍一个最强大的框架之一：微软.NET 框架。微软.NET 框架提供了许多优秀的编程工具（如，UI 窗体、模板、各种预定义控件、代码映射和 SQL Serve 开发工具），而且可以使用多种编程语言（如，C#、F#和 Visal Basic）进行开发。我们主要讨论 C++**通用语言运行库**（*Common Language Runtime*，CLI），也被称为 C++/CLI。CLI 表示通用语言运行库，是.NET 框架中的标准部分。

.NET 框架非常大型、复杂，我们将重点介绍多线程方案以及同步对象和技术。

## 6.2 托管代码和非托管代码

框架中包含各种各样的预处理控件和 UI 对象，可以把框架看作是带有丰富组件的一组例程和对象，让用户更容易地创建应用程序。我们编写的代码最终都要进行编译。编译后，通常会得到一个已编译的应用程序（*.exe）、一个动态链接库（*.dll）、一个内核驱动（*.sys）等。已编译的文件（或更多文件）中包含了由 0 和 1 组成的机器语言（代码），不方便我们阅读。根据定义，这样的代码叫非托管代码（*unmanaged code*）。本地 C++代码（即，我们编写的代码）被直接编译为机器语言，我们将其称为非托管代码。与托管代码不同，非托管代码运行速度快，直接在**硬件抽象层**（*Hardware Abstraction Layer*，HAL）上执行。HAL 提供了机器硬件和底层执行之间的直接交互。

Windows NT HAL 涉及直接处理计算机硬件的软件层。HAL 在硬件和 Windows NT 执行服务之间进行操作，应用程序和设备驱动不需要知道硬件的任何特定信息。HAL 隐藏了依赖硬件的细节（如，I/O 接口、中断控制器和多处理器通信机制）。应用程序和设备驱动必须调用 HAL 例程才能获知特定硬件信息，不允许直接处理硬件（摘自 MSDN）。

另一方面，托管代码稍有不同。它在通用语言运行库上执行，也就是说，在机器硬件和应用程序之间的某一层。你编写的代码不会被直接编译成机器语言，而是被编译成中间语言（MSIL，微软中间语言）。这对托管代码既有好处也有坏处。

我们先来分析有利的一面。托管代码被编译成 MSIL 后，还要进行各种错误检查，如静态类型检查（不在本地代码编译中执行）和运行时检查。托管代码提供了一个垃圾收集器（*garbage collector*），或者说托管代码能释放应用程序分配但尚未使用的动态内存等。.NET 框架提供了大量的编程工具，使用起来的确好处多多。

然而，有利就有弊。用任何框架（如，.NET 或 JRE[Java Runtime Environment]）开发应用程序都比用本地代码开发应用程序慢。因为托管代码在 CLR 层执行，这一层又在机器硬件层上执行。虽然用框架开发应用程序比较容易，但是它限制了对应用程序的控制和与机器之间的交互。框架提供什么，就只能做什么。

总而言之，如果希望应用程序执行得更快或完全控制应用程序的执行，就应该用本地方案来开发；如果希望开发有丰富 UI 的应用程序，且简化开发过程，就应该用框架。虽然多年以来，微软的.NET 框架日趋完善，但它不是万能的，本地方案仍有其用武之地。因此，本书只用一章的篇幅来介绍托管代码，其他章节都以讲解本地方案为主。

值得注意的是，用框架能完成的，用本地方案也能做到。只是后者实现起来费力一些，要花更多的时间和精力，不过的确可行。但是，一些本地方案能完成的事情，托管方案却未必可行。为了更好地理解这点，我们来看两个任务。第一个任务是创建一个控制核电站机器的软件，第二个任务是创建一个播放 DVD 的软件。第一个任务应该用本地 C++开发（或者甚至用汇编语言），以保障应用程序有极高的精确度和速度。原因无需多言，光提到核电两字都能让人神经紧绷，这可不是闹着玩的，必须要精确地掌控一切。相比之下，

DVD 软件的要求就没那么高，只要做到美观好用（这样可以热卖）即可，执行时间不需要精确到微秒或纳秒。作为一个 DVD 软件，开发的重点是界面友好且各种功能运行正常，所以用.NET 来开发既方便快捷又省时省力。

## 6.3 如何在.NET 中运行线程

运行在一个多核 CPU 上的应用程序，并发执行与逻辑核数量相同的并发任务才能最大程度地受益。当然，在执行并发任务时要认真考虑以下问题：

- 如何划分一组任务，让所有的任务都能在多核环境中执行？
- 如何确保并发任务的数量不超过 CPU 内核的数量？
- 如果一些任务停止了（如，在等待 I/O 时），如何发现这个状况并在等待 I/O 完成之前让 CUP 执行其他任务？
- 如何发现一个或多个并发操作执行完毕？

使用框架时，回答这些问题取决于框架本身。微软在 `System::Threading::Task` 名称空间中提供了 `Thread` 类、`Task` 类以及一系列辅助类型。

`Task` 类代表并发操作的抽象。CLR 使用 `Thread` 对象和 `ThreadPool` 类实现任务和调度执行任务。多线程应用程序中可以用 `Thread` 类执行，该类在 `System::Threading` 名称空间中。但是，如果要在 Windows 8 和 Windows 中储存应用程序，就不要用 `Thread` 类，应该用 `Task` 类。因为 `Task` 类提供了非常强大的多任务抽象。而且，如果设计的应用程序要创建指定数量（数量有限）的线程，那么该应用程序的扩展性会不够。也就是说，如果线程数量超过了内核数量，程序的反应就会变得迟钝。CLR 可以优化所需线程的数量，这样程序员就能根据可用内核的数量以最佳的方式完成并发任务。当应用程序创建一个 `Task` 对象时，一个实际任务就被添加到工作队列中。当线程可用时，任务将从工作队列中被移除，可用线程将执行该任务。为了更好地调度线程，`ThreadPool` 类实现了多种优化技术，使用了任务窃取（work-stealing）策略的一种算法。

> **任务窃取**是调度算法的一个特性，先分析问题再执行解决问题的线程。考虑时间片的数量和调用深度，如果问题很容易解决，就直接解决。如果问题相当复杂，为了解决问题，就用其他 CPU 内核上已创建、可用的线程将其划分成若干更小的部分。稍后，把子结果合并到最终的结果中。这里的关键是使用（窃取）可用的线程（如果有的话）。

接下来，我们将用 CLR 创建第一个示例。Visual Studio 在 2012 年和 2013 年前发布的版本中包含了 C++ Windows 窗体模板。2012 年和 2013 年发布的新版本不带窗体模板，不影响使用，只是用起来麻烦些。我们的第一个示例将创建一个简单的应用程序，计算 n 个迭代的阶乘（n 个或少于 n 个正整数的乘积）。工作线程进行计算，同时主线程更新进程条——只为了演示基本的 CLR 程序，并解释一个非常重要的特性：跨线程操作。

## 准备就绪

确定安装并运行了 Visual Studio。

## 操作步骤

现在，根据下面的步骤创建程序，稍后再详细解释。

1. 创建一个新的 C++ CLR 空项目，并命名为 `CLRApplication`。

2. 打开【解决方案资源管理器】，右键单击【源文件】，添加一个新的源文件 main。打开 main.cpp。输入下面的代码：

   ```cpp
   #include <Windows.h>
   #include <tchar.h>
   #include "MyForm.h"

   using namespace System::Windows::Forms;
   using namespace System::Threading;
   using namespace CLRApplication;

   [STAThreadAttribute]
   int APIENTRY _tWinMain(HINSTANCE hThis, HINSTANCE hPrev, LPSTR
       szCommandLine, int iCmdShow)
   {
       Application::EnableVisualStyles();
       Application::SetCompatibleTextRenderingDefault(false);

       Application::Run(gcnew MyForm());

       return 0;
   }
   ```

3. 打开【解决方案资源管理器】，右键单击【头文件】。在【Visial C++】下面选择 UI。添加一个 Windows 窗体 MyForm。打开 MyForm.h。

4. 输入下面的代码：

   ```cpp
   #pragma once
   namespace CLRApplication
   {
       using namespace System;
       using namespace System::Collections;
   ```

```cpp
using namespace System::Windows::Forms;
using namespace System::Drawing;
using namespace System::Threading::Tasks;

public ref class MyForm : public Form
{
public:
    MyForm(void)
        {
            InitializeComponent();
            this->Load += gcnew System::EventHandler(this,
                &MyForm::Form_OnLoad);
        }
protected:
    ~MyForm()
        {
            if (components)
            {
                delete components;
            }
        }
private:
    System::ComponentModel::Container ^components;
    Label^ label;
    ProgressBar^ progress;
    static const int N = 25;

    System::Void Form_OnLoad(System::Object^ sender,
        System::EventArgs^ e)
        {
            InitializeApp();
        }

    System::Void InitializeApp(void)
        {
            label = gcnew Label();
            label->Location = Point(2, 25);

            progress = gcnew ProgressBar();
            progress->Location = Point(2, 235);
            progress->Width = 280;
            progress->Height = 25;
            progress->Step = 1;
            progress->Maximum = MyForm::N;
            progress->Value = 0;

            this->Controls->Add(label);
            this->Controls->Add(progress);

            Task^ task = gcnew Task(gcnew Action<Object^>
                (&MyForm::StartAddress), this);
            task->Start();
        }

    static System::Void StartAddress(Object^ parameter)
        {
            MyForm^ form = (MyForm^)parameter;

            UpdateProgressBarDelegate^ action =
```

```
                gcnew UpdateProgressBarDelegate(form,
                    &MyForm::UpdateProgressBar);
            __int64 iResult = 1;

            for (int iIndex = 1; iIndex <= MyForm::N; iIndex++)
            {
                iResult *= iIndex;
                form->BeginInvoke(action);

                System::Threading::Thread::Sleep(100);
            }

            UpdateLabelTextDelegate^ textAction =
                gcnew UpdateLabelTextDelegate(form,
                    &MyForm::UpdateLabelText);
            form->BeginInvoke(textAction, L"Result: "
                + Convert::ToString(iResult));
        }

        delegate void UpdateProgressBarDelegate(void);

        System::Void UpdateProgressBar(void)
        {
            progress->PerformStep();
        }

        delegate void UpdateLabelTextDelegate(String^ szText);

        System::Void UpdateLabelText(String^ szText)
        {
            label->AutoSize = false;
            label->Text = szText;
            label->AutoSize = true;
        }

#pragma region Windows Form Designer generated code
        /// <summary>
        /// 设计者所需的方法
        /// 不要用代码编辑器修改该方法的内容
        /// </summary>
        void InitializeComponent(void)
        {
            this->AutoScaleMode =
                System::Windows::Forms::AutoScaleMode::Font;
            this->Size = System::Drawing::Size(300, 300);
            this->Text = L"MyForm";
            this->Padding = System::Windows::Forms::Padding(0);
            this->AutoScaleMode =
                System::Windows::Forms::AutoScaleMode::Font;
        }
#pragma endregion
    };
}
```

## 示例分析

首先解释 MyForm.h。每个 .NET 应用程序通常都从定义应用程序的名称空间开始。在上面的示例中，使用了 CLRApplication 默认名称空间（设计者制作的）。我们要使用框架的特性，所以需要指定所需的库（也叫做程序集），如下代码所示：

```
using namespace System;
using namespace System::Collections;
using namespace System::Windows::Forms;
using namespace System::Drawing;
using namespace System::Threading::Tasks;
```

有了这些程序集，我们才能使用 UI 工具。然后，派生出我们的 UI 类：MyForm，该类代表从 System::Windows::Forms::Form 基类派生的模板窗口。注意 MyForm 的构造函数，首先要调用 InitializeComponent，后面是我们的代码。InitializeComponent 载入一个组件的已编译页面，并初始化基础值，如大小、文本和内边距。InitializeComponent 的实现在一个叫做 Windows Form Designer generated code 的区域中。设计者把析构函数放入 MyForm 类的 protected 域中；在 private 域中有一个 components 变量，封装零个或多个组件（控件）。如下代码所示：

```
System::ComponentModel::Container ^components;
```

我们还添加了 label、prograss 和 N 变量分别表示进度条控件、标签控件和迭代次数。InitializeApp 方法用于创建一个新标签，并设置它的位置（在父控件上的 x 和 y 坐标）。除此之外，它还将创建一个新的进度条并初始化它的位置、大小（宽和高）、步数（调用 PerformStep 方法后进度条将更新多少步）、最大值（进度条的最大值）和开始值。然后，调用 Add 方法把这两个控件添加至主窗体：

```
this->Controls->Add( label );
this->Controls->Add( progress );
```

接下来创建线程（任务）。在本地方案中，我们必须指定线程的开始地址，如果需要的话，传递一个参数给它的开始例程。由于示例中的开始例程被标记为 static，所以为了稍后使用 MyForm 类，要把 this 指针传递给开始例程。任务构造函数需要一个指向 Action 对象的指针，如果要使用 Action<Object^> 重载，则还需要一个形参构造函数。我们将创建一个指向 Action<Object^>的指针，如下代码所示,：

```
gcnew Action<Object^>( &MyForm::StartAddress )
```

Action 构造函数需要委托（函数指针）来传递，我们提供的就是 StartAddress 方法的地址。

现在，我们来解释 StartAddress 方法。首先，要把 Object^（类似 void *）强制类型转换为 MyForm^。然后，实例化一个委托，稍后在调用主线程更新进度条时要用到。这部分非常重要，一定要理解！与本地方案不同，CLR 关注的是哪个线程创建了什么控件，只有创建了某个控件的线程才能操控该控件。因为主线程已经在 InitializeApp 方法中创建了进度条，所以只有主线程才能操控它。如果通过其他线程更新进度条，则会出现一个异常：**跨线程操作无效：只能从创建进度条的线程访问"进度"**。也就是说，工作线程必须请求主线程更新进度条，因为进度条是主线程创建的。这通过调用 Invoke 或 BeginInvoke 方法就能完成。这两个方法所做的工作相同，它们都在创建控件底层句柄（*underlying handle*）的线程上执行了一个委托。唯一的区别是，BeginInvoke 是异步执行，而 Invoke 是同步执行，要等待操作完成才返回。接下来，创建一个进行 n 次迭代的循环，在每次迭代的末尾调用 BeginInvoke 更新进度条。迭代完成后，将为标签文本执行相同的任务。

如前所述，我们要为前面描述的操作实例化委托。为了完成这项任务，必须声明委托和要调用的方法。如下代码所示：

```
delegate void UpdateProgressBarDelegate( void );
System::Void UpdateProgressBar( void )
{
    progress->PerformStep( );
}
```

我们将采用类似的步骤设置标签文本。创建好我们的类后，要指定应用程序的入口点。这在 main.cpp 中完成了。与非托管代码示例类似，我们要包含一些头文件才能使用预定义类型和 _tWinMain 方法。要包含 MyForm.h 窗体才能使用 MyForm 类和框架正常运行所需的程序集。为了使用 MyForm 类，还需要 System::Windows::Forms 程序集。为了把 MyForm 作为 Application::Run 方法的实例，还需要 System::Threading 才能设置线程套间状态和 CLRApplication 名称空间。我们将 ApartmentState 变量设置为 STA（**单线程套间**）。这样做是为了方便初始化为友好的 UI。例如，当我们要在 Windows UI 元素上使用拖放特性时，或者为了寄宿 COM 控件。Application::EnableVisualStyles 方法用于启动可视样式，丰富的用户界面风格，类似之前非托管示例中使用的 #pragma comment ( linker, ⋯ )。最后，调用 Application::Run 发布应用程序窗口并进入其消息循环。

## 更多讨论

我们已经讨论了这个.NET 应用程序的大部分内容，还有一些本书没有提到的部分，在此解释一下。.NET 框架包罗万象，需要看好几本书才能了解它的方方面面。我们着重解释那些对执行我们的应用程序影响最重要的部分，以及多线程概念必须提及的部分。其中一个要重点理解的是 gcnew 操作符。首先要明白，即使我们使用的是托管 C++，仍可以使用非托管代码。对于非托管代码，new 操作符用于分配有固定地址的内存，然后必须由程序员释放用过的内存。而托管代码与此不同，gcnew 操作符从.NET 框架托管的 CLR 堆中分配内存。这意味着分配的地址是不固定的，有可能在执行期间发生变化；也意味着程序员不必操心内存的回收问题。当已分配的内存不再使用时，CLR 会通过**垃圾收集器**（*garbage collector*，GC）回收内存。值得注意的是，^表示托管的指针，\*表示未托管的指针；%表示托管的引用，&表示非托管的引用。如表 6.1 所示。

表 6.1

| 操作符 | 非托管 | 托管 |
| --- | --- | --- |
| 间接引用（指针） | * | ^ |
| 引用 | & | % |

## 6.4 前台线程和后台线程的区别

.NET 有两种类型的线程：前台线程和后台线程。默认情况下，创建的线程都是前台线程。当然，也可以把线程状态显式设置为后台。两者的区别是，只要前台线程还在运行，应用程序就不会退出，而后台线程没这个特权。也就是说，关闭应用程序后，所有的后台线程都将自动终止。因此，当所有前台线程执行完毕时，应用程序可以在后台线程返回之前就退出。所有前台线程停止或退出后，系统将停止所有的后台线程。

上一节的程序示例中使用了 `Task` 对象，默认情况下该对象在后台线程中运行。本节的程序示例和上一节的示例基本相同，只是为了演示如何执行前台线程，用 `Thread` 对象代替了 `Task` 对象。

### 准备就绪

确定安装并运行了 Visual Studio。

### 操作步骤

现在，根据下面的步骤创建程序，稍后再详细解释。

1. 创建一个新的 C++ CLR 空项目，并命名为 `CLRApplication2`。

2. 打开【解决方案资源管理器】，右键单击【源文件】，添加一个新的源文件 main。打开 main.cpp。输入下面的代码：

   ```cpp
   #include <Windows.h>
   #include <tchar.h>
   #include "MyForm.h"

   using namespace System::Windows::Forms;
   using namespace System::Threading;
   using namespace CLRApplication2;

   [STAThreadAttribute]
   int APIENTRY _tWinMain(HINSTANCE hThis, HINSTANCE hPrev, LPTSTR
       szCommandLine, int iCmdShow)
   {
       Application::EnableVisualStyles();
       Application::SetCompatibleTextRenderingDefault(false);

       Application::Run(gcnew MyForm());

       return 0;
   }
   ```

3. 打开【解决方案资源管理器】，右键单击【头文件】。在【Visual C++】下选择 UI，添加一个新的 Windows 窗体，并命名为 `MyForm`。打开 MyForm.h。

4. 输入下面的代码：

   ```cpp
   #pragma once
   ```

```
namespace CLRApplication2
{
    using namespace System;
    using namespace System::Windows::Forms;
    using namespace System::Drawing;
    using namespace System::Threading;

    public ref class MyForm : public Form
    {
    public:
        MyForm(void)
        {
            InitializeComponent();
            this->Load += gcnew System::EventHandler(this,
                &MyForm::Form_OnLoad);
        }
    protected:
        ~MyForm()
        {
            if (components)
            {
                delete components;
            }
        }
    private:
        System::ComponentModel::Container ^components;
        Label^ label;
        ProgressBar^ progress;
        static const int N = 25;

        System::Void Form_OnLoad(System::Object^ sender,
            System::EventArgs^ e)
        {
            InitializeApp();
        }

        System::Void InitializeApp(void)
        {
            label = gcnew Label();
            label->Location = Point(2, 25);

            progress = gcnew ProgressBar();
            progress->Location = Point(2, 235);
            progress->Width = 280;
            progress->Height = 25;
            progress->Step = 1;
            progress->Maximum = MyForm::N;
            progress->Value = 0;

            this->Controls->Add(label);
            this->Controls->Add(progress);

            Thread^ thread = gcnew Thread(gcnew
                ParameterizedThreadStart(&MyForm::StartAddress));
            thread->Start(this);
        }

        static System::Void StartAddress(Object^ parameter)
        {
```

```
            MyForm^ form = (MyForm^)parameter;

            UpdateProgressBarDelegate^ action = gcnew
                UpdateProgressBarDelegate(form, &MyForm::UpdateProgressBar);

            __int64 iResult = 1;
            for (int iIndex = 1; iIndex <= MyForm::N; iIndex++)
            {
                iResult *= iIndex;
                form->BeginInvoke(action);

                System::Threading::Thread::Sleep(100);
            }

            UpdateLabelTextDelegate^ textAction = gcnew
                UpdateLabelTextDelegate(form, &MyForm::UpdateLabelText);
            form->BeginInvoke(textAction, L"Result: " +
                Convert::ToString(iResult));
        }

        delegate void UpdateProgressBarDelegate(void);

        System::Void UpdateProgressBar(void)
        {
            progress->PerformStep();
        }
        delegate void UpdateLabelTextDelegate(String^ szText);
        System::Void UpdateLabelText(String^ szText)
        {
            label->AutoSize = false;
            label->Text = szText;
            label->AutoSize = true;
        }

#pragma region Windows Form Designer generated code
        /// <summary>
        /// 设计者所需的方法
        /// 不要用代码编辑器修改该方法的内容
        /// </summary>
        void InitializeComponent(void)
        {
            this->components =
                gcnew System::ComponentModel::Container();
            this->Size = System::Drawing::Size(300, 300);
            this->Text = L"MyForm";
            this->Padding = System::Windows::Forms::Padding(0);
            this->AutoScaleMode =
                System::Windows::Forms::AutoScaleMode::Font;
        }
#pragma endregion
    };
}
```

## 示例分析

默认情况下,线程被创建为前台,而任务被创建为后台线程。在开发过程中,要根据具体情况选择合适的线程类型。例如,如果要计算一个耗时很长的大型阶乘(如,10000 的阶乘),就应该使用 Task 对象,这样只需关闭应用程序就能中断执行;如果程序必须在没有任何用户干扰的情况下完成它的任务(虽然有其他方式保证执行不会中断),就应该使用 Thread 对象,它能把任务作为前台线程执行。

### 更多讨论

还要指出一点,如果当前状况不能使用 Task 对象,那么 Thread 对象可以执行前台和后台两种类型。要把 Thread 作为后台线程,只需在启动该线程之前将其 IsBackground 属性改为 true 即可。如下代码所示:

```
Thread^ thread = gcnew Thread(
    gcnew ParameterizedThreadStart(&MyForm::StartAddress));
thread->IsBackground = true;
thread->Start(this);
```

## 6.5 理解.NET 同步要素

除了基本的线程操作(如,开始线程或任务)和提供线程的开始地址,.NET 还提供了一些同步机制。了解这些同步机制对正确理解线程的并发操作很有帮助。除此之外,我们还将解释阻塞方法、锁、触发和非阻塞同步。

在 .NET 中,阻塞方法有 Thread::Sleep、Thread::Join 和 Task::Wait。阻塞线程比自旋线程好,不会浪费处理器时间。这里提到的自旋是指循环,在满足某种条件之前线程将一直在循环中迭代。Sleep 方法把线程状态改为 suspended 一段时间。Join 方法在被调线程终止之前阻塞主调线程。Wait 方法在指定时间间隔内等待 Task 完成它的执行过程。

下面的示例中将计算 25 的阶乘,并询问用户是否继续。为了不浪费太多处理器时间,我们可以用 Sleep 方法作为最原始的自旋睡眠 1 秒。程序中还使用了 PostMessage API,以便在主线程和工作线程之间通信。作为最原始的通信方法,就目前而言足够了。稍后,我们将介绍其他通信技术。

### 准备就绪

确定安装并运行了 Visual Studio。

### 操作步骤

现在,根据下面的步骤创建程序,稍后再详细解释。

1. 创建一个新的 C++ CLR 空项目,并命名为 CLRApplication3。

2. 打开【解决方案资源管理器】,右键单击【源文件】,添加一个新的源文件 main。打开 main.cpp。输入下面的代码:

## 6.5 理解.NET 同步要素    189

```cpp
#include <Windows.h>
#include <tchar.h>
#include "MyForm.h"

using namespace System::Windows::Forms;
using namespace System::Threading;
using namespace CLRApplication3;

[STAThreadAttribute]
int APIENTRY _tWinMain(HINSTANCE hThis, HINSTANCE hPrev, LPTSTR
    szCommandLine, int iCmdShow)
{
    Application::EnableVisualStyles();
    Application::SetCompatibleTextRenderingDefault(false);

    Application::Run(gcnew MyForm());

    return 0;
}
```

3. 打开【解决方案资源管理器】,右键单击【头文件】。在【Visual C++】下选择 UI, 添加一个新的 Windows 窗体,并命名为 MyForm。打开 MyForm.h。

4. 输入下面的代码:

```cpp
#pragma once

#include <Windows.h>
#pragma comment ( lib, "User32.lib" )
#define END_TASK WM_USER + 1

namespace CLRApplication3
{
    using namespace System;
    using namespace System::Windows::Forms;
    using namespace System::Drawing;
    using namespace System::Threading::Tasks;

    public ref class MyForm : public Form
    {
    public:
    public:
        MyForm(void) : bContinue(true)
        {
            InitializeComponent();
            this->Load += gcnew System::EventHandler(this,
                &MyForm::Form_OnLoad);
        }
    protected:
        ~MyForm()
        {
            if (components)
            {
                delete components;
            }
        }
        virtual void WndProc(Message% msg) override
        {
            switch (msg.Msg)
```

```cpp
        {
            case END_TASK:
            {
                System::Windows::Forms::DialogResult^ dlgResult =
                    MessageBox::Show(L"Do you want to calculate again?",
                        L"Question", MessageBoxButtons::YesNo,
                        MessageBoxIcon::Question);
                if (*dlgResult ==
                    System::Windows::Forms::DialogResult::Yes)
                {
                    bContinue = true;
                    progress->Value = 0;
                    label->Text = L"Calculating...";
                }
                else
                {
                    PostMessage((HWND)this->Handle.ToPointer(), WM_CLOSE,
                        0, 0);
                }
                break;
            }
        }
        Form::WndProc(msg);
    }
private:
    System::ComponentModel::Container ^components;
    Label^ label;
    ProgressBar^ progress;
    static const int N = 25;
    bool bContinue;

    System::Void Form_OnLoad(System::Object^ sender,
        System::EventArgs^ e)
    {
        InitializeApp();
    }

    System::Void InitializeApp(void)
    {
        label = gcnew Label();
        label->Location = Point(2, 25);

        progress = gcnew ProgressBar();
        progress->Location = Point(2, 235);
        progress->Width = 280;
        progress->Height = 25;
        progress->Step = 1;
        progress->Maximum = MyForm::N;
        progress->Value = 0;

        this->Controls->Add(label);
        this->Controls->Add(progress);

        Task^ task = gcnew Task(gcnew Action<Object^>
            (&MyForm::StartAddress), this);
        task->Start();
    }

    static System::Void StartAddress(Object^ parameter)
```

```cpp
{
    MyForm^ form = (MyForm^)parameter;

    GetFormHandleDelegate^ getHandle = gcnew
        GetFormHandleDelegate(form, &MyForm::GetFormHandle);
    IntPtr handle = (IntPtr)form->Invoke(getHandle);

    UpdateProgressBarDelegate^ action = gcnew
        UpdateProgressBarDelegate(form, &MyForm::UpdateProgressBar);
    UpdateLabelTextDelegate^ textAction = gcnew
        UpdateLabelTextDelegate(form, &MyForm::UpdateLabelText);

    Int64 iResult = 1;

    while (true)
    {
        if (!form->bContinue)
        {
            System::Threading::Thread::Sleep(1000);
            continue;
        }

        iResult = 1;
        for (int iIndex = 1; iIndex <= MyForm::N; iIndex++)
        {
            iResult *= iIndex;
            form->BeginInvoke(action);

            System::Threading::Thread::Sleep(100);
        }

        form->BeginInvoke(textAction, L"Result: " +
            Convert::ToString(iResult));

        PostMessage((HWND)handle.ToPointer(), END_TASK, 0, 0);

        form->bContinue = false;

        System::Threading::Thread::Sleep(1000);
    }
}

delegate void UpdateProgressBarDelegate(void);
System::Void UpdateProgressBar(void)
{
    progress->PerformStep();
}

delegate void UpdateLabelTextDelegate(String^ szText);
System::Void UpdateLabelText(String^ szText)
{
    label->AutoSize = false;
    label->Text = szText;
    label->AutoSize = true;
}

delegate IntPtr GetFormHandleDelegate(void);
System::IntPtr GetFormHandle(void)
{
```

```
                return this->Handle;
        }
#pragma region Windows Form Designer generated code
        /// <summary>
        /// 设计者所需的方法
        /// 不要用代码编辑器修改该方法的内容
        /// </summary>
        void InitializeComponent(void)
        {
            this->components =
                gcnew System::ComponentModel::Container();

            this->Size = System::Drawing::Size(300, 300);
            this->Text = L"MyForm";
            this->Padding = System::Windows::Forms::Padding(0);
            this->AutoScaleMode =
                System::Windows::Forms::AutoScaleMode::Font;
        }
#pragma endregion
    };
}
```

## 示例分析

该示例与前面的示例几乎相同，只是为了在后台运行线程，我们选择了 `Task` 对象。因为当用户选择不继续时，应用程序将退出，工作线程将被强制退出。当然，我们也可以选择 `Thread` 对象并设置 `IsBackground` 属性。由于要使用 `PostMessage API`，所以要包含 `User32` 库，如下代码所示：

```
#pragma comment (lib, "User32.lib")
```

如果不使用框架，默认情况下会包含 `User32`（及其他）库，但是不会包含 `mscorlib.dll`。而使用框架时，默认情况下会包含 `mscorlib.dll` 库，但是不会包含 `User32`。接着，为了能在线程间通信，我们定义了 `END_TASK` 消息。我们将扩展我们的类，把 `WndProc` 重载方法添加到主窗口的进程消息中。再次强调，我们使用窗口消息队列来触发主线程。此外，还必须添加另一个代理。该代理将调用 `GetFormHandle` 方法并获得主窗口的句柄，工作线程要给该窗口寄送消息。我们还改变了 `StartAddress` 方法，为了不浪费太多处理器时间，现在线程在 1 秒钟睡眠间隔中执行一个无限循环。每次计算完成后，将触发主线程询问用户是否继续运行应用程序。如果用户选择否，将简单地寄送 `WM_CLOSE` 消息，并终止后台线程（即，`Task` 对象），并退出应用程序。然而，如果用户选择继续，我们将再次执行计算。

## 更多讨论

本节演示了用最原始的技术来完成给定的任务。我们用 `PostMessage` 触发主线程，用 `Sleep` 方法在等待用户回答时自旋线程。在后面的示例中，我们将演示一些.NET 特性，利用这些特性通过集成（或内置）对象和技术来完成同样的任务。

## 6.6 锁和避免死锁

在上一节的示例中，两个线程都可以对 bContinue 变量进行读写操作，没有任何限制。这通常为应用程序埋下了安全隐患，应该避免。微软.NET 框架提供了一些机制来避免类似的情况。

下面的示例将实现一个 Lock 类，与第 3 章的 CLock 类相似。Lock 类将帮助我们在另一个线程正在写的时候停止要读取的线程，反之亦然。C#语言提供了 lock 操作符，但是 C++没有。因此，我们要实现自己的锁，使用 Monitor 对象锁住线程的执行，并在其他线程访问共享对象时等待。

### 准备就绪

确定安装并运行了 Visual Studio。

### 操作步骤

现在，根据下面的步骤创建程序，稍后再详细解释。

1. 创建一个新的 C++ CLR 空项目，并命名为 CLRApplication4。

2. 打开【解决方案资源管理器】，右键单击【源文件】，添加一个新的源文件 main。打开 main.cpp。输入下面的代码：

    ```
    #include <Windows.h>
    #include <tchar.h>
    #include "MyForm.h"

    using namespace System::Windows::Forms;
    using namespace System::Threading;
    using namespace CLRApplication4;

    [STAThreadAttribute]
    int APIENTRY _tWinMain(HINSTANCE hThis, HINSTANCE hPrev, LPTSTR
        szCommandLine, int iCmdShow)
    {
        Application::EnableVisualStyles();
        Application::SetCompatibleTextRenderingDefault(false);

        Application::Run(gcnew MyForm());

        return 0;
    }
    ```

3. 打开【解决方案资源管理器】，右键单击【头文件】。在【Visual C++】下选择 UI，添加一个新的 Windows 窗体，并命名为 MyForm。打开 MyForm.h。

4. 输入下面的代码：

    ```
    #pragma once
    ```

```cpp
#include <Windows.h>
#pragma comment ( lib, "User32.lib" )

#define END_TASK WM_USER + 1

namespace CLRApplication4
{
    using namespace System;
    using namespace System::Windows::Forms;
    using namespace System::Drawing;
    using namespace System::Threading;
    using namespace System::Threading::Tasks;

    ref class Lock
    {
        Object^ lockObject;
    public:
        Lock(Object^ lock) : lockObject(lock)
        {
            Monitor::Enter(lockObject);
        }
    protected:
        ~Lock()
        {
            Monitor::Exit(lockObject);
        }
    };

    public ref class MyForm : public Form
    {
    public:
        MyForm(void) : bContinue(true)
        {
            lockObject = gcnew Object();
            InitializeComponent();
            this->Load += gcnew System::EventHandler(this,
                &MyForm::Form_OnLoad);
        }
    protected:
        ~MyForm()
        {
            if (components)
            {
                delete components;
            }
            delete lockObject;
        }

        virtual void WndProc(Message% msg) override
        {
            switch (msg.Msg)
            {
                case END_TASK:
                {
                    System::Windows::Forms::DialogResult^ dlgResult =
                        MessageBox::Show(L"Do you want to calculate again?",
                            L"Question", MessageBoxButtons::YesNo,
                            MessageBoxIcon::Question);
```

```cpp
                    if (*dlgResult ==
                        System::Windows::Forms::DialogResult::Yes)
                    {
                        SetState(true);
                        progress->Value = 0;
                        label->Text = L"Calculating...";
                    }
                    else
                    {
                        PostMessage((HWND)this->Handle.ToPointer(), WM_CLOSE,
                            0, 0);
                    }
                    break;
                }
            }
            Form::WndProc(msg);
        }
private:
        System::ComponentModel::Container ^components;
        Label^ label;
        ProgressBar^ progress;
        static const int N = 25;
        bool bContinue;
        Object^ lockObject;

        System::Void Form_OnLoad(System::Object^ sender,
            System::EventArgs^ e)
        {
            InitializeApp();
        }

        System::Void SetState(bool bState)
        {
            Lock^ lock = gcnew Lock(this->lockObject);
            bContinue = bState;
            delete lock;
        }

        bool QueryState(void)
        {
            Lock^ lock = gcnew Lock(this->lockObject);
            bool bState = bContinue;
            delete lock;
            return bState;
        }

        System::Void InitializeApp(void)
        {
            label = gcnew Label();
            label->Location = Point(2, 25);

            progress = gcnew ProgressBar();
            progress->Location = Point(2, 235);
            progress->Width = 280;
            progress->Height = 25;
            progress->Step = 1;
            progress->Maximum = MyForm::N;
            progress->Value = 0;
```

```cpp
            this->Controls->Add(label);
            this->Controls->Add(progress);

            Task^ task = gcnew Task(gcnew Action<Object^>
                (&MyForm::StartAddress), this);
            task->Start();
        }

        static System::Void StartAddress(Object^ parameter)
        {
            MyForm^ form = (MyForm^)parameter;

            GetFormHandleDelegate^ getHandle = gcnew
                GetFormHandleDelegate(form, &MyForm::GetFormHandle);
            IntPtr handle = (IntPtr)form->Invoke(getHandle);

            UpdateProgressBarDelegate^ action = gcnew
                UpdateProgressBarDelegate(form, &MyForm::UpdateProgressBar);
            UpdateLabelTextDelegate^ textAction = gcnew
                UpdateLabelTextDelegate(form, &MyForm::UpdateLabelText);

            Int64 iResult = 1;
            while (true)
            {
                if (!form->QueryState())
                {
                    Thread::Sleep(1000);
                    continue;
                }

                iResult = 1;

                for (int iIndex = 1; iIndex <= MyForm::N; iIndex++)
                {
                    iResult *= iIndex;
                    form->BeginInvoke(action);
                    Thread::Sleep(100);
                }

                form->BeginInvoke(textAction, L"Result: " +
                    Convert::ToString(iResult));

                PostMessage((HWND)handle.ToPointer(), END_TASK, 0, 0);

                form->bContinue = false;

                Thread::Sleep(1000);
            }
        }

        delegate void UpdateProgressBarDelegate(void);
        System::Void UpdateProgressBar(void)
        {
            progress->PerformStep();
        }

        delegate void UpdateLabelTextDelegate(String^ szText);
        System::Void UpdateLabelText(String^ szText)
```

```
        {
            label->AutoSize = false;
            label->Text = szText;
            label->AutoSize = true;
        }
        delegate IntPtr GetFormHandleDelegate(void);
        System::IntPtr GetFormHandle(void)
        {
            return this->Handle;
        }
#pragma region Windows Form Designer generated code
        /// <summary>
        /// 设计者所需的方法
        /// 不要用代码编辑器修改该方法的内容
        /// </summary>
        void InitializeComponent(void)
        {
            this->components =
                gcnew System::ComponentModel::Container();
            this->Size = System::Drawing::Size(300, 300);
            this->Text = L"MyForm";
            this->Padding = System::Windows::Forms::Padding(0);
            this->AutoScaleMode =
                System::Windows::Forms::AutoScaleMode::Font;
        }
#pragma endregion
    };
}
```

## 示例分析

首先，我们添加了一个新的 Lock 类。Lock 对象背后的思想是，防止共享对象同时被多个线程访问。为此，要使用 System::Threading 名称空间中的 Monitor 对象以及 Enter 方法和 Exit 方法。要使用 Monitor::Enter 和 Monitor::Exit，必须提供指向 System::Object 的指针，以防止共享对象同时被多个线程访问。在访问共享对象之前，实例化指向 Lock 对象的指针，并在访问结束后立即删除该指针。因此，要在 Lock 析构函数中调用 Monitor::Exit。这样，我们就能安全地设置 bContinue 的值或读取它，不用担心是否有多个线程同时访问它，导致不可预知的结果。

## 更多讨论

接下来，该讨论死锁了。死锁是一种情况，多个线程正在等待独占访问某个对象，但是持有该对象的线程已被挂起或某种原因不能继续执行。当线程 A 正在等待线程 B，而线程 B 也在等待线程 A 时，就会出现死锁。如图 6.1 所示。

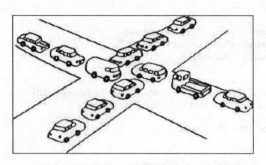

图 6.1 死锁图示

在独占访问的许多情况下都会出现死锁。例如，我们在程序中添加了独占访问，所以主线程在设置 bContinue 的值时，工作线程不能读取它，或者在工作线程读取它的值时，主线程不能写入。如果一锁定 bContinue 对象就挂起工作线程会怎样？我们会和死锁不期而遇。因为主线程在工作线程继续工作之前被阻塞，而工作线程又被挂起不能继续工作了。检测、避免和成功处理死锁的算法很多，但是这超出了本书讨论的范围。第 8 章将详细介绍如何检测和处理死锁。在操作系统领域，不乏深入研究死锁理论的详实材料，这里是为了提醒开发者哪些情况下可能出现死锁。

## 6.7 线程安全和.NET 框架的类型

什么是线程安全？线程安全指的是在没有干扰的前提下，单独线程能执行独占访问的能力。为什么在使用.NET 时线程安全很重要？这源于.NET 某些类型和操作的内部实现，微软把它们设计成线程安全的。以相同的方式继续开发是良好的编程习惯（也就是说，编写线程安全的代码），可以避免在设计应用程序编码时产生问题。

在.NET 中，所有的静态成员都是线程安全的，而实例成员不是。例如，如果两个线程要并发枚举某个 List 集合，它们会分别获得不同的枚举数对象。如果它们在枚举中都不修改集合值就没问题。但是，如果要在枚举期间修改，就必须使用锁。

如下代码所示，通过列表进行枚举是线程安全的：

```
List<int>^ list = gcnew List<int>();
for (int iValue = 0; iValue < 5; iValue++)
{
    list->Add(iValue);
}
array<Object^>^ _t1Params = gcnew array<Object^>(2)
{
    list, L"Task _t1"
};
Task^ _t1 = gcnew Task(gcnew Action<Object^>(&EnumerateList), _t1Params);
```

```
    _t1->Start();

    array<Object^>^ _t2Params = gcnew array<Object^>(2)
    {
        list, L"Task _t2"
    };
    Task^ _t2 = gcnew Task(gcnew Action<Object^>(&EnumerateList), _t2Params);
    _t2->Start();
}
static void EnumerateList(Object^ parameter)
{
    array<Object^>^ _tParams = (array<Object^>^) parameter;

    List<int>^ lst = (List<int>^) _tParams[0];

    String^ szValues = gcnew String(L" ");

    for each (int iValue in lst)
    {
        szValues += iValue + L" ";
    }

    MessageBox::Show(szValues, (String^)_tParams[1]);
}
...
```

如前所述，每个线程都将获得不同的枚举对象。因此，可以说线程是安全的。.NET 提供的另一个特性是 initonly 关键字，它能安全地并发读操作。当有两个（或多个）线程访问这种类型时，是安全的。如下代码所示：

```
ref class SafeValue
{
public:
    SafeValue(void) { }
    void Run(void)
    {
        Task^ _t1 = gcnew Task(gcnew Action(&AccessValue));
        _t1->Start();

        Task^ _t2 = gcnew Task(gcnew Action(&AccessValue));
        _t2->Start();
    }
protected:
    ~SafeValue(void) { }
private:
    static void AccessValue(void)
    {
        MessageBox::Show(Convert::ToString(SafeValue::iValue));
    }
    static initonly int iValue = 10;
};

int main()
{
    SafeValue^ safeValue = gcnew SafeValue();
    safeValue->Run();
}
```

在应用程序中添加安全，意味着所有独占访问的地方都要用锁上下围起来，但是同时，你应该意识到独占意味着线程要等待，结果是花费大量的时间，可能导致潜在的死锁。再次强调，在开始实际编码之前，一定要好好设计应用程序。第 7 章将详细介绍如何思考和设计一个并行应用程序，避免所有潜在的危险状况和发现任务本身的问题。

## 6.8 事件等待句柄的触发

第 6.5 节的程序示例中使用了不安全的方式访问 bContinue 共享对象，还调用了 Sleep 方法等待用户选择是否继续运行应用程序。稍后，第 6.6 节中的程序示例添加了锁确保独占访问，把应用程序改进为线程安全的。当然，循环中仍然调用了 Sleep 方法。微软.NET 框架提供了一些触发对象（*signaling object*），与前几章介绍的 Event 很类似。这些触发对象是 AutoResetEvent、ManualResetEvent、ManualResetEventSlim、CountDownEvent、Barrier、Wait 和 Pulse 方法。

下面的程序示例使用 ManualResetEventSlim，因为它的开销非常小，只花费大约 40 纳秒。我们移除了 Lock 类和访问共享对象，不需要再使用它们了。当工作线程第一次计算完毕时，将通知主线程：操作已完成，并调用 ManualResetEventSlim::Wait 方法挂起线程的执行。稍后，如果用户选择继续计算，只需设置事件对象就能触发线程继续执行。

### 准备就绪

确定安装并运行了 **Visual Studio**。

### 操作步骤

现在，根据下面的步骤创建程序，稍后再详细解释。

1. 创建一个新的 **C++ CLR** 空项目，并命名为 CLRApplication5。

2. 打开【解决方案资源管理器】，右键单击【源文件】，添加一个新的源文件 main。打开 main.cpp。输入下面的代码：

```
#include <Windows.h>
#include <tchar.h>
#include "MyForm.h"

using namespace System::Windows::Forms;
using namespace System::Threading;
using namespace CLRApplication5;

[STAThreadAttribute]
int APIENTRY _tWinMain(HINSTANCE hThis, HINSTANCE hPrev, LPTSTR
    szCommandLine, int iCmdShow)
{
```

```
        Application::EnableVisualStyles();
        Application::SetCompatibleTextRenderingDefault(false);

        Application::Run(gcnew MyForm());

        return 0;
    }
```

3. 打开【解决方案资源管理器】，右键单击【头文件】。在【Visual C++】下选择 UI，添加一个新的 Windows 窗体，并命名为 MyForm。打开 MyForm.h。

4. 输入下面的代码：

```
#pragma once

#include <Windows.h>
#pragma comment ( lib, "User32.lib" )

#define END_TASK WM_USER + 1

namespace CLRApplication5
{
    using namespace System;
    using namespace System::Windows::Forms;
    using namespace System::Drawing;
    using namespace System::Threading;
    using namespace System::Threading::Tasks;

    public ref class MyForm : public Form
    {
    public:
        MyForm(void)
        {
            mEvent = gcnew ManualResetEventSlim(false);
            InitializeComponent();
            this->Load += gcnew System::EventHandler(this,
                &MyForm::Form_OnLoad);
        }
    protected:
        ~MyForm()
        {
            if (components)
            {
                delete components;
            }
            delete mEvent;
        }

        virtual void WndProc(Message% msg) override
        {
            switch (msg.Msg)
            {
                case END_TASK:
                {
                    System::Windows::Forms::DialogResult^ dlgResult =
                        MessageBox::Show(L"Do you want to calculate again?",
                            L"Question", MessageBoxButtons::YesNo,
                            MessageBoxIcon::Question);
```

```cpp
                    if (*dlgResult ==
                        System::Windows::Forms::DialogResult::Yes)
                    {
                        progress->Value = 0;
                        label->Text = L"Calculating...";
                        mEvent->Set();
                    }
                    else
                    {
                        this->Close();
                    }
                    break;
                }
            }

            Form::WndProc(msg);
        }
    private:
        System::ComponentModel::Container ^components;
        Label^ label;
        ProgressBar^ progress;
        static const int N = 25;
        static ManualResetEventSlim^ mEvent;

        System::Void Form_OnLoad(System::Object^ sender,
            System::EventArgs^ e)
        {
            InitializeApp();
        }

        System::Void InitializeApp(void)
        {
            label = gcnew Label();
            label->Location = Point(2, 25);

            progress = gcnew ProgressBar();
            progress->Location = Point(2, 235);
            progress->Width = 280;
            progress->Height = 25;
            progress->Step = 1;
            progress->Maximum = MyForm::N;
            progress->Value = 0;

            this->Controls->Add(label);
            this->Controls->Add(progress);

            Task^ task = gcnew Task(gcnew Action<Object^>
                (&MyForm::StartAddress), this);
            task->Start();
        }

        static System::Void StartAddress(Object^ parameter)
        {
            MyForm^ form = (MyForm^)parameter;

            GetFormHandleDelegate^ getHandle = gcnew
                GetFormHandleDelegate(form, &MyForm::GetFormHandle);
            IntPtr handle = (IntPtr)form->Invoke(getHandle);
```

```
            UpdateProgressBarDelegate^ action = gcnew
                UpdateProgressBarDelegate(form, &MyForm::UpdateProgressBar);
            UpdateLabelTextDelegate^ textAction = gcnew
                UpdateLabelTextDelegate(form, &MyForm::UpdateLabelText);

            Int64 iResult = 1;
            while (true)
            {
                iResult = 1;
                mEvent->Reset();

                for (int iIndex = 1; iIndex <= MyForm::N; iIndex++)
                {
                    iResult *= iIndex;
                    form->BeginInvoke(action);

                    Sleep(100);
                }

                form->BeginInvoke(textAction, L"Result: " +
                    Convert::ToString(iResult));

                PostMessage((HWND)handle.ToPointer(), END_TASK, 0, 0);

                mEvent->Wait();
            }
        }

        delegate void UpdateProgressBarDelegate(void);
        System::Void UpdateProgressBar(void)
        {
            progress->PerformStep();
        }

        delegate void UpdateLabelTextDelegate(String^ szText);
        System::Void UpdateLabelText(String^ szText)
        {
            label->AutoSize = false;
            label->Text = szText;
            label->AutoSize = true;
        }

        delegate IntPtr GetFormHandleDelegate(void);
        System::IntPtr GetFormHandle(void)
        {
            return this->Handle;
        }

#pragma region Windows Form Designer generated code
        /// <summary>
        /// 设计者所需的方法
        /// 不要用代码编辑器修改该方法的内容
        /// </summary>
        void InitializeComponent(void)
        {
            this->components = gcnew
                System::ComponentModel::Container();
            this->Size = System::Drawing::Size(300, 300);
```

```cpp
                this->Text = L"MyForm";
                this->Padding = 
                    System::Windows::Forms::Padding(0);
                this->AutoScaleMode = 
                    System::Windows::Forms::AutoScaleMode::Font;
            }
#pragma endregion
        };
    }
```

## 示例分析

该示例比前两个示例的解决方案好。这次，我们移除了 `Lock` 类，同时不再使用共享对象（这是出问题的根源）。我们在 `private` 域中添加了 `ManualResetEventSlim` 变量，出于安全考虑，声明为 `static`。在 `MyForm` 的构造函数中，将创建一个新的 `ManualResetEventSlim` 实例。为简单起见，该程序示例只考虑一个实例的情况。多个实例的情况留给读者来完成。另外，`ManualResetEventSlim` 的构造函数需要一个 `bool` 类型的参数，用于设置事件的初始状态是 `signaled` 还是 `nonsignaled`。工作线程开始执行时，将把事件的状态设置为 `nonsignaled`。通过调用 `Wait` 方法，将进入挂起状态，等待事件变成已触发状态。在主线程的 `WndProc` 函数中，用户选择继续后，时间状态将被设置为 `signaled`。这样，工作线程就能继续执行了。

## 更多讨论

如果要求等待时间短，且不能在进程间共享事件，那么用 `ManualResetEventSlim` 比用 `ManualResetEvent` 更快速，性能更好。通常情况下，等待时间越短，`ManualResetEventSlim` 的开销越微不足道，除非等待某些等待句柄。然而，如果事件状态不能在短时间内变成 `signaled`，事件的行为就表现得和常规的 `ManualResetEvent` 对象差不多了。

# 6.9 基于事件的异步模式

微软.NET 提供了**基于事件的异步模式**（*Event-based Asynchronous Pattern*，EAP），让程序员不用显式创建或启动一个线程或任务就能使用多线程。这些类是 `BackgroundWorker` 或 `WebClient`。稍后，将重点介绍 `BackgroundWorker` 类。现在，先简要介绍一下 `WebClient` 的特性，并修改我们的阶乘示例，用事件代替寄送消息。

托管 `WebClient` 类及其他类提供了同步和异步下载数据的方法。我们感兴趣的是异步方法。代码如下所示：

```
[HostProtectionAttribute(SecurityAction::LinkDemand, ExternalThreading = true)]
public: void DownloadDataAsync( Uri^ address )
public: void DownloadDataAsync( Uri^ address, Object^ userToken )
public: event DownloadDataCompletedEventHandler^ DownloadDataCompleted
{
```

```cpp
        void add (DownloadDataCompletedEventHandler^ value);
        void remove (DownloadDataCompletedEventHandler^ value);
}
```

异步方法背后的思想是，这些方法创建一个新的线程执行下载操作，而开发者要注册事件通知。或者换言之，它必须提供一个在操作完成时调用的方法。

我们要在阶乘示例中做这样的改动。为了询问用户是否重复操作，我们要创建一个计算结束就触发的事件。虽然之前的示例中，通过发布消息给主线程的窗口消息队列已经完成了这项任务，但是不能总这样做。如果创建的是控制台应用程序怎么办？没有窗口，就不能使用窗口消息队列了。因此，我们要详细了解事件的机制，这很重要。

## 准备就绪

确定安装并运行了 Visual Studio。

## 操作步骤

现在，根据下面的步骤创建程序，稍后再详细解释。

1. 创建一个新的 C++ CLR 空项目，并命名为 `CLRApplication6`。

2. 打开【解决方案资源管理器】，右键单击【源文件】，添加一个新的源文件 main。打开 main.cpp。输入下面的代码：

    ```cpp
    #include <Windows.h>
    #include <tchar.h>
    #include "MyForm.h"

    using namespace System::Windows::Forms;
    using namespace System::Threading;
    using namespace CLRApplication6;

    [STAThreadAttribute]
    int APIENTRY _tWinMain(HINSTANCE hThis, HINSTANCE hPrev, LPTSTR
        szCommandLine, int iCmdShow)
    {
        Application::EnableVisualStyles();
        Application::SetCompatibleTextRenderingDefault(false);

        Application::Run(gcnew MyForm());

        return 0;
    }
    ```

3. 打开【解决方案资源管理器】，右键单击【头文件】。在【Visual C++】下选择 UI，添加一个新的 Windows 窗体，并命名为 `MyForm`。打开 MyForm.h。

4. 输入下面的代码：

```cpp
#pragma once

namespace CLRApplication6
{
    using namespace System;
    using namespace System::Windows::Forms;
    using namespace System::Drawing;
    using namespace System::Threading;
    using namespace System::Threading::Tasks;

    public ref class MyForm : public Form
    {
    public:
        MyForm(void)
        {
            InitializeComponent();
            mEvent = gcnew ManualResetEventSlim(false);
            EndTaskEvent += gcnew EndTaskEventHandler(&MyForm::EndTask);
            updateProgressBar = gcnew UpdateProgressBarDelegate(this,
                &MyForm::UpdateProgressBar);
            updateLabelText = gcnew UpdateLabelTextDelegate(this,
                &MyForm::UpdateLabelText);
            updateProgressBarValue = gcnew
                UpdateProgressBarValueDelegate(this,
                    &MyForm::UpdateProgressBarValue);
            closeForm = gcnew FormCloseDelegate(this,
                &MyForm::FormClose);

            this->Load += gcnew System::EventHandler(this,
                &MyForm::Form_OnLoad);
        }
    protected:
        ~MyForm()
        {
            if (components)
            {
                delete components;
            }
            delete mEvent;
        }
    private:
        System::ComponentModel::Container ^components;

        delegate void UpdateProgressBarDelegate(void);
        System::Void UpdateProgressBar(void)
        {
            progress->PerformStep();
        }

        delegate void UpdateProgressBarValueDelegate(int iValue);
        System::Void UpdateProgressBarValue(int iValue)
        {
            progress->Value = iValue;
        }
        delegate void UpdateLabelTextDelegate(String^ szText);
        System::Void UpdateLabelText(String^ szText)
        {
            label->AutoSize = false;
            label->Text = szText;
```

```cpp
        label->AutoSize = true;
}

delegate void FormCloseDelegate(void);
System::Void FormClose(void)
{
        this->Close();
}

Label^ label;
ProgressBar^ progress;
static const int N = 25;
static ManualResetEventSlim^ mEvent;
delegate void EndTaskEventHandler(MyForm^);
static event EndTaskEventHandler^ EndTaskEvent;
UpdateProgressBarDelegate^ updateProgressBar;
UpdateLabelTextDelegate^ updateLabelText;
UpdateProgressBarValueDelegate^ updateProgressBarValue;
FormCloseDelegate^ closeForm;

System::Void Form_OnLoad(Object^ sender, System::EventArgs^ e)
{
        InitializeApp();
}

static System::Void EndTask(MyForm^ form)
{
        System::Windows::Forms::DialogResult^ dlgResult =
            MessageBox::Show(L"Do you want to calculate again?", L"Question",
                MessageBoxButtons::YesNo, MessageBoxIcon::Question);

        if (*dlgResult == System::Windows::Forms::DialogResult::Yes)
        {
            if (form->progress->InvokeRequired)
            {
                form->BeginInvoke(form->updateProgressBarValue,
                    (Object^)0);
            }
            else
            {
                form->progress->Value = 0;
            }
            if (form->label->InvokeRequired)
            {
                form->BeginInvoke(form->updateLabelText,
                    L"Calculating...");
            }
            else
            {
                form->label->Text = L"Calculating...";
            }

            mEvent->Set();
        }
        else
        {
            if (form->InvokeRequired)
            {
                form->BeginInvoke(form->closeForm);
```

```cpp
            }
            else
            {
                form->Close();
            }
        }

        System::Void InitializeApp(void)
        {
            label = gcnew Label();
            label->Location = Point(2, 25);

            progress = gcnew ProgressBar();
            progress->Location = Point(2, 235);
            progress->Width = 280;
            progress->Height = 25;
            progress->Step = 1;
            progress->Maximum = MyForm::N;
            progress->Value = 0;

            this->Controls->Add(label);
            this->Controls->Add(progress);

            Task^ task = gcnew Task(gcnew Action<Object^>
                (&MyForm::StartAddress), this);
            task->Start();
        }

        static System::Void StartAddress(Object^ parameter)
        {
            MyForm^ form = (MyForm^)parameter;
            Int64 iResult = 1;

            while (true)
            {
                iResult = 1;

                mEvent->Reset();

                for (int iIndex = 1; iIndex <= MyForm::N; iIndex++)
                {
                    iResult *= iIndex;
                    form->BeginInvoke(form->updateProgressBar);
                    System::Threading::Thread::Sleep(100);
                }

                form->BeginInvoke(form->updateLabelText, L"Result: " +
                    Convert::ToString(iResult));

                MyForm::EndTaskEvent(form);

                mEvent->Wait();
            }
        }

#pragma region Windows Form Designer generated code
        /// <summary>
        /// 设计者所需的方法,
```

```
            /// 不要用代码编辑器修改该方法的内容
            /// </summary>
            void InitializeComponent(void)
            {
                this->components = gcnew
                    System::ComponentModel::Container();
                this->Size = System::Drawing::Size(300, 300);
                this->Text = L"MyForm";
                this->Padding =
                    System::Windows::Forms::Padding(0);
                this->AutoScaleMode =
                    System::Windows::Forms::AutoScaleMode::Font;
            }
    #pragma endregion
        };
    }
```

## 示例分析

本例演示了事件的用法，移除了给主窗口消息队列寄送消息。这样做的好处是，应用程序不再依赖窗口。不过，也有一个坏处。由于工作线程触发事件（调用一个例程），所以必须调用合适的方法才能用 UI 元素进行操控（如，改变进度条或标签文本的值）。我们故意创建一个这样的例子，是为了强调跨线程操作。在 .NET 中进行任何异步操作，都会影响 UI 控件的操控。理解这一点非常重要。接下来，分析一下代码。我们改变了 MyForm 的构造函数，前 6 行代码创建新的事件和委托实例；把委托从局部声明的变量改成类特征（因为要供多个方法使用）；添加了 static 例程 EndTask，该例程将在引发 EndTaskEventHandler 事件时被调用；最后，引发一个事件询问用户是否继续计算。如此一来，我们采用了一种稍微不同的方案，以正确的方式使用了 .NET 框架提供的 EAP。

## 更多讨论

为了让读者理解 EAP 的构建，下面演示了 WebClientDownloadDataAsync 方法的用法：

```
...
String^ szUrl = L"http://cplusplus.expert.its.me/testfile.txt";
WebClient^ client = gcnew WebClient();
client->DownloadDataCompleted +=
gcnew DownloadDataCompletedEventHandler(
    &MyForm::AsyncDownloadDataCompleted);
client->DownloadDataAsync(gcnew Uri(szUrl));
...
static void AsyncDownloadDataCompleted(Object^ sender,
    DownloadDataCompletedEventArgs^ e)
{
    String^ szResult;
    System::Text::Encoding^ enc = System::Text::Encoding::UTF8;
    szResult = enc->GetString(e->Result);
    if (e->Error == nullptr
        && !String::IsNullOrEmpty(szResult))
    {
        System::IO::File::WriteAllText(
            L"C:\\testFile.txt", szResult);
        MessageBox::Show(L"Download commplete.");
```

          }
    }

与上一个示例类似，这里要注册一个事件，我们用 `AsyncDownloadDataCompleted` 方法来完成。在调用 `WebClient::DownloadDataAsync` 时，CLR 创建了一个新线程，将并行地执行下载操作。执行完毕后，将引发一个随后调用 `AsyncDownloadDataCompleted` 方法的事件。

## 6.10　BackgoundWorker 类

上一节的示例说明了 EAP 的机制非常好，程序员不用编写太多代码就能轻松地创建多线程应用程序。`BackgoundWorker` 对象是 EAP 的通用实现，有许多特性，如方便取消模型、安全地更新另一个线程的 UI 控件、报告进度特性、转发异常、引发完成事件等。下面的示例中，将使用 `System::ComponentModel` 名称空间中的 `BackgroundWorker` 类实现我们的阶乘示例。

### 准备就绪

确定安装并运行了 Visual Studio。

### 操作步骤

现在，根据下面的步骤创建程序，稍后再详细解释。

1. 创建一个新的 C++ CLR 空项目，并命名为 `CLRApplication7`。

2. 打开【解决方案资源管理器】，右键单击【源文件】，添加一个新的源文件 main。打开 main.cpp。输入下面的代码：

    ```cpp
    #include <Windows.h>
    #include <tchar.h>
    #include "MyForm.h"

    using namespace System::Windows::Forms;
    using namespace System::Threading;
    using namespace CLRApplication7;

    [STAThreadAttribute]
    int APIENTRY _tWinMain(HINSTANCE hThis, HINSTANCE hPrev, LPTSTR
        szCommandLine, int iCmdShow)
    {
        Application::EnableVisualStyles();
        Application::SetCompatibleTextRenderingDefault(false);

        Application::Run(gcnew MyForm());

        return 0;
    }
    ```

3. 打开【解决方案资源管理器】,右键单击【头文件】。在【Visual C++】下选择 UI,添加一个新的 Windows 窗体,并命名为 MyForm。打开 MyForm.h。

4. 输入下面的代码:

```cpp
#pragma once

namespace CLRApplication7
{
    using namespace System;
    using namespace System::ComponentModel;
    using namespace System::Windows::Forms;
    using namespace System::Drawing;

    public ref class MyForm : public Form
    {
    public:
        MyForm(void)
        {
            InitializeComponent();

            this->Load += gcnew System::EventHandler(this,
                &MyForm::Form_OnLoad);
        }
    protected:
        ~MyForm()
        {
            if (components)
            {
                delete components;
            }
        }
    private:
        System::ComponentModel::Container ^components;
        Label^ label;
        ProgressBar^ progress;

        static const int N = 25;

        System::Void Form_OnLoad(System::Object^ sender,
            System::EventArgs^ e)
        {
            InitializeApp();
        }

        System::Void InitializeApp(void)
        {
            label = gcnew Label();
            label->Location = Point(2, 25);

            progress = gcnew ProgressBar();
            progress->Location = Point(2, 235);
            progress->Width = 280;
            progress->Height = 25;
            progress->Step = 1;
            progress->Maximum = MyForm::N;
            progress->Value = 0;
```

```
                this->Controls->Add(label);
                this->Controls->Add(progress);

                BackgroundWorker^ bckWorker = gcnew BackgroundWorker();
                bckWorker->WorkerReportsProgress = true;
                bckWorker->ProgressChanged +=
                    gcnew ProgressChangedEventHandler(this,
                        &MyForm::ProgressChanged);

                bckWorker->RunWorkerCompleted +=
                    gcnew RunWorkerCompletedEventHandler(this,
                        &MyForm::RunningCompleted);
                bckWorker->DoWork += gcnew DoWorkEventHandler(this,
                    &MyForm::StartAddress);
                bckWorker->RunWorkerAsync(bckWorker);
            }

            System::Void StartAddress(Object^ sender, DoWorkEventArgs^ e)
            {
                BackgroundWorker^ bckWorker = (BackgroundWorker^)e->Argument;

                Int64 iResult = 1;
                for (int iIndex = 1; iIndex <= MyForm::N; iIndex++)
                {
                    iResult *= iIndex;

                    bckWorker->ReportProgress(iIndex);

                    System::Threading::Thread::Sleep(100);
                }

                e->Result = iResult;
            }

            System::Void ProgressChanged(Object^ sender,
                ProgressChangedEventArgs^ e)
            {
                progress->Value = e->ProgressPercentage;
            }

            System::Void RunningCompleted(Object^ sender,
                RunWorkerCompletedEventArgs^ e)
            {
                label->AutoSize = false;
                label->Text = L"Result: " + Convert::ToString(e->Result);
                label->AutoSize = true;
            }

#pragma region Windows Form Designer generated code
            /// <summary>
            /// 设计者所需的方法,
            /// 不要用代码编辑器修改该方法的内容
            /// </summary>
            void InitializeComponent(void)
            {
                this->components = gcnew
                    System::ComponentModel::Container();
                this->Size = System::Drawing::Size(300, 300);
```

```
                    this->Text = L"MyForm";
                    this->Padding =
                        System::Windows::Forms::Padding(0);
                    this->AutoScaleMode =
                        System::Windows::Forms::AutoScaleMode::Font;
                }
#pragma endregion
            };
        }
```

## 示例分析

读者应该注意到了，我们越深入地了解框架的特性，开发时间就越短，需要编码的数量就越少。在前几个版本的阶乘应用程序中，必须实现委托才能操控不同的线程 UI 控件。BackgroundWorker 中的集成事件（如，ProgressChanged 或 RunWorkerCompleted）都帮我们搞定了。BackgroundWorker 的这些特性减轻了编码的负担。跨线程经常是程序员的噩梦，现在再也不用担心跨线程操作了。而且，编程逻辑也更清晰明了。我们来分析一下代码。主要的改动在 InitializeApp 中，必须正确地初始化 BackgroundWorker 类对象；必须为 3 个事件提供方法委托，并把 WorkerReportsProgress 正确设置为 true。注意，为了匹配事件的原型，必须改变 StartAddress 原型，如下所示：

```
System::Void StartAddress(Object^ sender, DoWorkEventArgs^ e)
```

我们仍然可以通过 RunWorkerAsync 方法传递一个参数，稍后在 StartAddress 方法中通过 DoWorkEventArgs 类对象的 Argument 属性获得它。在需要引发 ProgressChanged 事件时，只需调用 BackgroundWorker::ReportProgress 方法并传递一个整数作为参数即可，该参数指定了更新进度的次数。最后，为了知道何时工作线程执行完毕，我们注册了 RunWorkerCompleted 事件。当线程退出时，就引发事件。

## 更多讨论

Thread 类是作为 sealed 类来实现的，而 BackgroundWorker 类与它不同。这意味着 BackgroundWorker 类可以有自己的子类，而且可以根据所需的情况从 BackgroundWorker 对象派生出其他版本。当你要处理一个复杂的任务时（如，天气预报），这很方便。BackgroundWorker 类应该有许多属性，以保存待处理的大量数据。注意，如果从 BackgroundWorker 类派生，还要使用 OnDoWork 虚函数。类似下面这样：

```
ref class Worker : BackgroundWorker
{
public:
    Worker() : BackgroundWorker()
    {
        WorkerReportsProgress = true;
    }
protected:
    virtual void OnDoWork(DoWorkEventArgs^ e) override
    {
```

```
            Worker^ bckWorker = (Worker^)e->Argument;
            Int64 iResult = 1;
            for (int iIndex = 1; iIndex <= Worker::N; iIndex++)
            {
                iResult *= iIndex;
                bckWorker->ReportProgress(iIndex);
                System::Threading::Thread::Sleep(100);
            }
            e->Result = iResult;
        }
    private:
        static const int N = 25;
        // 添加字段
    };
```

其用法如下：

```
Worker^ worker = gcnew Worker();
worker->RunWorkerAsync(worker);
```

## 6.11 中断、中止和安全取消线程执行

一些任务要强制中断线程的执行。在前面的示例中，我们使用了 Task 和 BackgoundWorker 对象，主要是要用到它们的后台操作，而且它们能在应用程序退出时强制关闭。接下来介绍一些用于常规线程中断和安全取消的.NET 特性。如果未满足某些解除阻塞的条件或者未设置超时，阻塞方法将永远阻塞线程。然而，有时要在继续执行应用程序的前提下取消线程的执行。在其他情况下，要取消线程（例如，接收用户输入），然后再次开启线程。结束线程的方法是 Thread::Interrupt 和 Thread::Abort。

Interrupt 方法不会中断正在运行的线程。调用 Interrupt 方法时，在抛出 ThreadInterruptedException 时，未被阻塞的线程在下次被阻塞之前将继续执行；而在被阻塞的线程上，则调用中断强制终止该线程，并再次抛出 ThreadInterruptedException。Abort 方法才会对非阻塞线程起作用，请记住几乎总是应该用 Abort 代替 Interrupt。当使用 Abort 方法时，会抛出类似的异常：ThreadAbortException。尽管如此，Abort 和 Interrupt 方法都被认为是不安全的。因为如果在线程例程中有一个非托管代码执行，当你使用 Abort 时，线程将继续运行到下一个执行托管代码的语句为止。另一方面，Interrupt 在线程未被阻塞之前不会影响线程的执行。

在下面的程序示例中，将演示如何用 Abort 方法中止一个线程。

### 准备就绪

确定安装并运行了 Visual Studio。

### 操作步骤

现在，根据下面的步骤创建程序，稍后再详细解释。

1. 创建一个新的 C++ CLR 空项目，并命名为 CLRApplication8。

2. 打开【解决方案资源管理器】，右键单击【源文件】，添加一个新的源文件 main。打开 main.cpp。
   输入下面的代码：

   ```
   #include <Windows.h>
   #include <tchar.h>
   #include "MyForm.h"

   using namespace System::Windows::Forms;
   using namespace System::Threading;
   using namespace CLRApplication8;

   [STAThreadAttribute]
   int APIENTRY _tWinMain(HINSTANCE hThis, HINSTANCE hPrev, LPTSTR
       szCommandLine, int iCmdShow)
   {
       Application::EnableVisualStyles();
       Application::SetCompatibleTextRenderingDefault(false);

       Application::Run(gcnew MyForm());

       return 0;
   }
   ```

3. 打开【解决方案资源管理器】，右键单击【头文件】。在【Visual C++】下选择 UI，添加一个新的 Windows 窗体，并命名为 MyForm。打开 MyForm.h。

4. 输入下面的代码：

   ```
   #pragma once

   namespace CLRApplication8
   {
       using namespace System;
       using namespace System::Threading;
       using namespace System::Windows::Forms;
       using namespace System::Drawing;

       public ref class MyForm : public Form
       {
       public:
           MyForm(void)
           {
               InitializeComponent();
               this->Load += gcnew System::EventHandler(this,
                   &MyForm::Form_OnLoad);

               updateProgressBar = gcnew UpdateProgressBarDelegate(this,
                   &MyForm::UpdateProgressBar);
               updateLabelText = gcnew UpdateLabelTextDelegate(this,
                   &MyForm::UpdateLabelText);
               this->FormClosing += gcnew FormClosingEventHandler(this,
   ```

```cpp
                    &MyForm::FormClosingEvent);
        }
    protected:
        ~MyForm()
        {
            if (components)
            {
                delete components;
            }
        }
    private:
        delegate void UpdateProgressBarDelegate(void);
        System::Void UpdateProgressBar(void)
        {
            progress->PerformStep();
        }

        delegate void UpdateLabelTextDelegate(String^ szText);
        System::Void UpdateLabelText(String^ szText)
        {
            label->AutoSize = false;
            label->Text = szText;
            label->AutoSize = true;
        }

        System::ComponentModel::Container ^components;
        Label^ label;
        ProgressBar^ progress;
        static const int N = 25;
        Thread^ worker;
        UpdateProgressBarDelegate^ updateProgressBar;
        UpdateLabelTextDelegate^ updateLabelText;

        System::Void Form_OnLoad(System::Object^ sender,
            System::EventArgs^ e)
        {
            InitializeApp();
        }

        System::Void FormClosingEvent(Object^ sender,
            FormClosingEventArgs^ e)
        {
            System::Windows::Forms::DialogResult^ dlgResult =
                MessageBox::Show(L"Do you want to abort execution? ",
                    L"Question", MessageBoxButtons::YesNo,
                    MessageBoxIcon::Question);

            if (*dlgResult == System::Windows::Forms::DialogResult::Yes)
            {
                worker->Abort();
            }
        }

        System::Void InitializeApp(void)
        {
            label = gcnew Label();
            label->Location = Point(2, 25);

            progress = gcnew ProgressBar();
```

```
                progress->Location = Point(2, 235);
                progress->Width = 280;
                progress->Height = 25;
                progress->Step = 1;
                progress->Maximum = MyForm::N;
                progress->Value = 0;

                this->Controls->Add(label);
                this->Controls->Add(progress);

                worker = gcnew Thread(gcnew ParameterizedThreadStart(
                    &MyForm::StartAddress));
                worker->Start(this);
            }

            static System::Void StartAddress(Object^ parameter)
            {
                MyForm^ form = (MyForm^)parameter;

                Int64 iResult = 1;
                for (int iIndex = 1; iIndex <= MyForm::N; iIndex++)
                {
                    iResult *= iIndex;
                    form->BeginInvoke(form->updateProgressBar);

                    Thread::Sleep(100);
                }

                form->BeginInvoke(form->updateLabelText, L"Result: " +
                    Convert::ToString(iResult));
            }

#pragma region Windows Form Designer generated code
            /// <summary>
            /// 设计者所需的方法,
            /// 不要用代码编辑器修改该方法的内容
            /// </summary>
            void InitializeComponent(void)
            {
                this->SuspendLayout();
                this->AutoScaleDimensions =
                    System::Drawing::SizeF(6, 13);
                this->AutoScaleMode =
                    System::Windows::Forms::AutoScaleMode::Font;
                this->Size = System::Drawing::Size(284, 261);
                this->MaximizeBox = false;
                this->MinimizeBox = false;
                this->Name = L"MyForm";
                this->Text = L"MyForm";
                this->ResumeLayout(false);
            }
#pragma endregion
    };
}
```

**示例分析**

我们仍然使用相同的示例。该示例仅为了向读者演示如何使用 Abort 方法。我们提醒读者要谨慎使用这些方法，只有在自己的线程实现中才知道这样做会发生什么。

## 安全取消

虽然 .NET 框架实现了 Abort 和 Interrupt 方法，但还是要尽量避免使用。这里我们推荐使用另一种方式：BackgroundWorker::WorkerSupportsCancellation 属性和它的 CancelAsync 方法。

下面的示例演示了如何用 CancelAsync 方法中止 BackgroundWorker 线程。

### 准备就绪

确定安装并运行了 **Visual Studio**。

### 操作步骤

现在，根据下面的步骤创建程序，稍后再详细解释。

1. 创建一个新的 **C++ CLR** 空项目，并命名为 CLRApplication9。

2. 打开【解决方案资源管理器】，右键单击【源文件】，添加一个新的源文件 main。打开 main.cpp。输入下面的代码：

    ```cpp
    #include <Windows.h>
    #include <tchar.h>
    #include "MyForm.h"

    using namespace System::Windows::Forms;
    using namespace System::Threading;
    using namespace CLRApplication9;

    [STAThreadAttribute]
    int APIENTRY _tWinMain(HINSTANCE hThis, HINSTANCE hPrev, LPTSTR
        szCommandLine, int iCmdShow)
    {
        Application::EnableVisualStyles();
        Application::SetCompatibleTextRenderingDefault(false);

        Application::Run(gcnew MyForm());

        return 0;
    }
    ```

3. 打开【解决方案资源管理器】，右键单击【头文件】。在【Visual C++】下选择 UI，添加一个新的 Windows 窗体，并命名为 MyForm。打开 MyForm.h。

4. 输入下面的代码：

```cpp
#pragma once

namespace CLRApplication9
{
    using namespace System;
    using namespace System::ComponentModel;
    using namespace System::Windows::Forms;
    using namespace System::Drawing;

    public ref class MyForm : public Form
    {
    public:
        MyForm(void)
        {
            InitializeComponent();
            this->Load += gcnew System::EventHandler(this,
                &MyForm::Form_OnLoad);

            this->FormClosing += gcnew FormClosingEventHandler(this,
                &MyForm::FormClosingEvent);
        }
    protected:
        ~MyForm()
        {
            if (components)
            {
                delete components;
            }
        }
    private:
        System::ComponentModel::Container ^components;
        Label^ label;
        ProgressBar^ progress;
        static const int N = 25;

        BackgroundWorker^ bckWorker;

        System::Void Form_OnLoad(Object^ sender, EventArgs^ e)
        {
            InitializeApp();
        }

        System::Void FormClosingEvent(Object^ sender,
            FormClosingEventArgs^ e)
        {
            System::Windows::Forms::DialogResult^ dlgResult =
                MessageBox::Show(L"Do you want to abort execution?", L"Question",
                    MessageBoxButtons::YesNo, MessageBoxIcon::Question);

            if (*dlgResult == System::Windows::Forms::DialogResult::Yes)
            {
                bckWorker->CancelAsync();
            }
            else
            {
                e->Cancel = true;
            }
        }
```

```
System::Void InitializeApp(void)
{
    label = gcnew Label();
    label->Location = Point(2, 25);

    progress = gcnew ProgressBar();
    progress->Location = Point(2, 235);
    progress->Width = 280;
    progress->Height = 25;
    progress->Step = 1;
    progress->Maximum = MyForm::N;
    progress->Value = 0;

    this->Controls->Add(label);
    this->Controls->Add(progress);

    bckWorker = gcnew BackgroundWorker();
    bckWorker->WorkerReportsProgress = true;
    bckWorker->WorkerSupportsCancellation = true;
    bckWorker->ProgressChanged +=
        gcnew ProgressChangedEventHandler(this,
            &MyForm::ProgressChanged);
    bckWorker->RunWorkerCompleted +=
        gcnew RunWorkerCompletedEventHandler(this,
            &MyForm::RunningCompleted);
    bckWorker->DoWork += gcnew DoWorkEventHandler(this,
        &MyForm::StartAddress);
    bckWorker->RunWorkerAsync(bckWorker);
}

System::Void StartAddress(Object^ sender, DoWorkEventArgs^ e)
{
    BackgroundWorker^ worker = (BackgroundWorker^)e->Argument;
    Int64 iResult = 1;
    for (int iIndex = 1; iIndex <= MyForm::N; iIndex++)
    {
        iResult *= iIndex;
        worker->ReportProgress(iIndex);

        System::Threading::Thread::Sleep(100);

        if (worker->CancellationPending)
        {
            return;
        }
    }

    e->Result = iResult;
}

System::Void ProgressChanged(Object^ sender,
    ProgressChangedEventArgs^ e)
{
    progress->Value = e->ProgressPercentage;
}

System::Void RunningCompleted(Object^ sender,
    RunWorkerCompletedEventArgs^ e)
{
```

```
            label->AutoSize = false;
            label->Text = L"Result: " + Convert::ToString(e->Result);
            label->AutoSize = true;
        }

#pragma region Windows Form Designer generated code
        /// <summary>
        /// 设计者所需的方法，
        /// 不要用代码编辑器修改该方法的内容
        /// </summary>
        void InitializeComponent(void)
        {
            this->components = gcnew
                System::ComponentModel::Container();
            this->Size = System::Drawing::Size(300, 300);
            this->Text = L"MyForm";
            this->Padding =
                System::Windows::Forms::Padding(0);
            this->AutoScaleMode =
                System::Windows::Forms::AutoScaleMode::Font;
        }
#pragma endregion
    };
}
```

## 示例分析

我们完成了和上一个示例相同的任务，但是这次是安全的，使用了 .NET 的其他集成特性：BackgroundWorker::CancelAsync 方法和 BackgroundWorker 的属性 WorkerSupportsCancellation。但是，在一些特殊的情况中，特别是必须完全控制线程执行，而且必须立即执行取消线程的情况下，不能这样实现。此时，应该使用一个 bool 类型（或其他类型）的变量模仿相应的属性，然后用 Lock 对象将其上下围起来，以确保独占访问。

回头看一下代码。我们添加了 FormClosingEventHandler 来处理 form closing 事件，把 BackgroundWorker 属性 WorkerSupportsCancellation 设置为 true。在 StartAddress 里面，添加了下面的语句：

```
if (worker->CancellationPending)
{
    return;
}
```

如果需要的话，以上代码以一种有序的方式取消线程的执行。

## 更多讨论

最后，来分析 Task 类。.NET 的 Task 对象提供了一个非常好用的新特性：ContinueWith 方法。在当前任务结束时，可以使用延续（*continuation*）立刻为下一个任务做好准备。

对于那些只在某任务执行完毕后才执行的任务而言，延续很好用。例如，假设有两个任务，第 2 个任务必须使用第 1 个任务的结果,那么就应该考虑把第 2 个任务作为延续任务。当使用 ContinueWith 方法时,分发器重新分配线程,用提供给 ContinueWith 方法作为参数的开始地址继续执行。如下代码所示:

```
static System::Void StartAddress(Object^ parameter);
static System::Void ContinueAddress(Task^ task);
...
Task^ task = gcnew Task(gcnew Action<Object^>(&StartAddress), this);
task->Start( );
Task^ continueTask = task->ContinueWith(gcnew Action<Task^>( &ContinueAddress));
...
```

ContinueWith 方法有多个重载,可以使用不同的值,如 TaskContinuationOptions、TaskScheduler 或 TaskCreationOptions。欲详细了解相关内容,请查阅 MSDN(http://msdn.microsoft.com/en-us/library/system.threading.tasks.taskcontinuationoptions%28v=vs.110%29.aspx?cs-save-lang=1&cs-lang=cpp#code-snippet-1)。在众多好处中,Task 对象提供了 WaitAll 和 WaitAny 方法,与 WaitForSingleObject 和 WaitForMultipleObjects API 类似。

## 6.12 非阻塞同步

在 .NET 提供的众多特性中,程序员必须意识到某些 CLR 行为才能避免在使用多线程时犯错。为了提高效率或优化缓存,微软 CLI/C++ 编译器和 CLR 都能在运行时重新排序某些语句。由于某些原因,这样做并不一定总能很好地组织代码。有时必须把一些语句保护起来。因此,.NET 提供了 MemoryBarrier 类防止指令重新排序的效果和读/写缓存,如下面的类所示:

```
...
ref class TestClass
{
public:
    TestClass()
    {
        bState = false;
        iValue = 0;
    }
    static void SetMethod()
    {
        iValue = 5;
        bState = true;
    }
    static void GetMethod()
    {
        if (bState)
        {
            MessageBox::Show(Convert::ToString(iValue));
        }
    }
private:
    static bool bState;
    static int iValue;
};
```

```
ref class TestWorker
{
public:
    TestWorker()
    {
        TestClass _t;
        Task^ _t1 = gcnew Task(gcnew Action(&_t.SetMethod));
        _t1->Start();
        Task^ _t2 = gcnew Task(gcnew Action(&_t.GetMethod));
        _t2->Start();
    }
};
...
```

读者是否认为 GetMethod 中的 iValue 应该为 0？根据前面提到的内容，的确如此。最简单的解决方案是创建内存栅栏（*memory fence*）或者用 MemoryBarrier 语句把 bState 特征上下围起来，如下代码所示：

```
...
static void SetMethod()
{
    iValue = 5;
    Thread::MemoryBarrier();
    bState = true;
    Thread::MemoryBarrier();
}

static void GetMethod()
{
    Thread::MemoryBarrier();
    if (bState)
    {
        Thread::MemoryBarrier();
        MessageBox::Show(Convert::ToString(iValue));
    }
}
...
```

这叫做完全栅栏（*full fence*），执行这种内存保护大约需要 10 纳秒的时间。

C++编译器还提供了另一种特性：volatile 关键字。使用 volatile 关键字，命令编译器为每个读操作在 volatile 内存区域创建一个获取栅栏（*acquire fence*），在声明为 volatile 的内存区域为每个写操作创建一个释放栅栏（*release fence*）。获取栅栏防止读/写操作被移动至栅栏上面，释放栅栏防止读/写操作被移动至栅栏下面。用 volatile 内存区域来实现，如下代码所示：

```
...
ref class TestClass
{
public:
    TestClass()
    {
        bState = false;
        iValue = 0;
```

```
        }
        static void SetMethod()
        {
            iValue = 5;
            bState = true;
        }
        static void GetMethod()
        {
            if (bState)
            {
                MessageBox::Show(Convert::ToString(iValue));
            }
        }
    private:
        static volatile bool bState;
        static int iValue;
    };
    ...
```

由于给运行时优化留有更多的空间,用 volatile 内存对象或所谓的半栅栏（*half fence*）更快速。

## 6.13 Wait 和 Pulse 触发

在前面的示例中,Lock 类的实现中用到了 Monitor::Enter 和 Monitor::Exit。Monitor 对象的另一个非常重要的特性是 Monitor::Wait 和 Monitor::Pulse 方法（以及 Pulse 的变式 PulseAll）。这些方法也构成了 EAP。它们等待接收其他线程信号的能力非常好用。最重要的特性之一是它们防止自旋,处理器的时间可浪费不起。唯一的限制是,必须用锁（Monitor::Enter 和 Monitor::Exit）把它们上下围起来。再次强调,我们引入锁是为了独占访问共享对象。

下面的示例演示了如何把静态 Wait 和 Pulse 方法作为一种同步多线程的技术。

### 准备就绪

确定安装并运行了 Visual Studio。

### 操作步骤

现在,根据下面的步骤创建程序,稍后再详细解释。

1. 创建一个新的 C++ CLR 空项目,并命名为 CLRApplication10。

2. 打开【解决方案资源管理器】,右键单击【源文件】,添加一个新的源文件 main。打开 main.cpp。
   输入下面的代码:

   ```
   #include <Windows.h>
   #include <tchar.h>
   #include "MyForm.h"
   ```

```
using namespace System::Windows::Forms;
using namespace System::Threading;
using namespace CLRApplication10;

[STAThreadAttribute]
int APIENTRY _tWinMain(HINSTANCE hThis, HINSTANCE hPrev, LPTSTR
    szCommandLine, int iCmdShow)
{
    Application::EnableVisualStyles();
    Application::SetCompatibleTextRenderingDefault(false);

    Application::Run(gcnew MyForm());

    return 0;
}
```

3. 打开【解决方案资源管理器】，右键单击【头文件】。在【Visual C++】下选择 UI，添加一个新的 Windows 窗体，并命名为 MyForm。打开 MyForm.h。

4. 输入下面的代码：

```
#pragma once

namespace CLRApplication10
{
    using namespace System;
    using namespace System::Threading;
    using namespace System::Threading::Tasks;
    using namespace System::Windows::Forms;
    using namespace System::Drawing;

    ref class Lock
    {
        Object^ lockObject;
    public:
        Lock(Object^ lock) : lockObject(lock)
        {
            Monitor::Enter(lockObject);
        }
    protected:
        ~Lock()
        {
            Monitor::Exit(lockObject);
        }
    };

    public ref class MyForm : public Form
    {
    public:
        MyForm(void) : syncHandle(gcnew Object())
        {
            InitializeComponent();
            this->Load += gcnew EventHandler(this,
                &MyForm::Form_OnLoad);
        }
    protected:
        ~MyForm()
```

```
            {
                if (components)
                {
                    delete components;
                }
            }
            virtual void SetWait(void)
            {
                Lock^ lock = gcnew Lock(syncHandle);
                Monitor::Wait(syncHandle);
                delete lock;
            }
            virtual void SetPulse(void)
            {
                Lock^ lock = gcnew Lock(syncHandle);
                Monitor::Pulse(syncHandle);
                delete lock;
            }
        private:
            System::ComponentModel::Container ^components;
            Label^ label;
            ProgressBar^ progress;
            static const unsigned N = 25;
            initonly Object^ syncHandle;

            System::Void Form_OnLoad(Object^ sender, EventArgs^ e)
            {
                InitializeApp();
            }

            delegate void TaskCompletedDelegate(UInt64 uResult);
            void TaskCompleted(UInt64 uResult)
            {
                label->AutoSize = false;
                label->Text = L"Result: " + uResult;
                label->AutoSize = true;

                System::Windows::Forms::DialogResult^ dlgResult =
                    MessageBox::Show(L"Do you want to continue?", L"Question",
                        MessageBoxButtons::YesNo, MessageBoxIcon::Question);

                if (*dlgResult == System::Windows::Forms::DialogResult::Yes)
                {
                    this->SetPulse();
                    progress->Value = 0;
                    label->Text = L"Calculating...";
                }
                else
                {
                    this->Close();
                }
            }

            delegate void UpdateProgressBarDelegate(void);
            System::Void UpdateProgressBar(void)
            {
                progress->PerformStep();
            }
```

```cpp
System::Void InitializeApp(void)
{
    label = gcnew Label();
    label->Location = Point(2, 25);

    progress = gcnew ProgressBar();
    progress->Location = Point(2, 235);
    progress->Width = 280;
    progress->Height = 25;
    progress->Step = 1;
    progress->Maximum = MyForm::N;
    progress->Value = 0;

    this->Controls->Add(label);
    this->Controls->Add(progress);

    Task^ task = gcnew Task(gcnew Action<Object^>
        (&MyForm::StartAddress), gcnew array<Object^>(2)
    {
        this, task
    });
    task->Start();
}

static System::Void StartAddress(Object^ parameter)
{
    array<Object^>^ parameters = (array<Object^>^) parameter;

    MyForm^ form = (MyForm^)parameters[0];
    Task^ task = (Task^)parameters[1];

    TaskCompletedDelegate^ taskCompleted = gcnew
        TaskCompletedDelegate(form, &MyForm::TaskCompleted);
    UpdateProgressBarDelegate^ updateProgressBar = gcnew
        UpdateProgressBarDelegate(form, &MyForm::UpdateProgressBar);

    UInt64 uResult = 1;
    while (true)
    {
        uResult = 1;

        for (unsigned uIndex = 1; uIndex <= MyForm::N; uIndex++)
        {
            uResult *= uIndex;
            form->progress->BeginInvoke(updateProgressBar);
            Thread::Sleep(100);
        }
        form->BeginInvoke(taskCompleted, uResult);

        form->SetWait();
    }
}

#pragma region Windows Form Designer generated code
    /// <summary>
    /// 设计者所需的方法,
    /// 不要用代码编辑器修改该方法的内容
    /// </summary>
    void InitializeComponent(void)
```

```
            {
                this->SuspendLayout();
                this->AutoScaleDimensions =
                    System::Drawing::SizeF(6, 13);
                this->AutoScaleMode =
                    System::Windows::Forms::AutoScaleMode::Font;
                this->Size = System::Drawing::Size(284, 300);
                this->MaximizeBox = false;
                this->MinimizeBox = false;
                this->Name = L"MyForm";
                this->Text = L"MyForm";
                this->ResumeLayout(false);
            }
#pragma endregion
    };
}
```

### 示例分析

Lock 类的实现没有变化，我们在 MyForm 类中增加了新方法：SetWait 和 SetPulse。这些方法声明在 protected 中，主要是为了让其他程序员可以从 MyForm 类派生以扩展行为。前面提到过，程序员必须用锁把 Wait 和 Pulse 上下围起来。在 private 域中，我们用 initonly syncHandle 特征扩展了 MyForm 类。该特征将作为所有 Monitor 方法调用的同步对象使用。与 StartAddress 方法不同的是，我们使用了 Control::BeginInvoke 方法异步调用 TaskCompleted 例程。原因是，TaskCompleted 方法操控主线程创建的 UI，如果我们在工作线程中同步调用 TaskCompleted 方法，就要为 TaskCompleted 内部所有的语句都创建委托。这当然会引发跨线程操作的异常。而异步调用 TaskCompleted，就不用编写这些代码，而且把控制权交给了主线程。我们使用 Monitor::Wait 方法在用户回答是否继续任务之前等待。然后，为了不浪费处理器的时间，工作线程进入挂起状态。既然控制权已经交给了主线程，那么现在的任务是询问用户是否继续计算。如果用户回答"是"，则调用 Monitor::Pulse 把工作线程的状态改成 nonsignaled。然后继续执行工作线程，重复计算过程。

### 更多讨论

Monitor::Wait 和 Monitor::Pulse，再配合锁，能提供非常好的性能，需要写的代码也很少。ResetEvent（自动或手动）或 Semaphore 也能提供相同的特性完成相同的任务。只不过，如果知道执行 Pulse 只需大约 100 纳秒，而 Set 方法触发事件句柄要大约 400 纳秒，你一定会尽量考虑用 Pulse。

## 6.14 Barrier 类

在用于协调任务的同步基本类中，.NET 的特性 Barrier 类表现突出。虽然锁机制非常好用，但是在有些非常复杂的任务中，总有些棘手的问题让你抓狂。Barrier 类能让你暂时停止（暂停）执行应用程序中的一个任务，或者在某处暂停任务集，当所有任务都到达时再继续执行。为了并行地执行一系列多任务，这个特性在同步方面发挥着重要的作用。

当应用程序创建一个 Barrier 对象时,它必须在将被同步的 set 中指定任务的数量。Barrier 对象内部的任务计数器会用到这个值。用 Barrier::AddParticipant 方法可以递增该值,用 Barrier::RemoveParticipant 方法可以递减该值。当任务到达同步点时,将调用 Barrier::SignalAndAwait 方法,在 Barrier 对象内部递减线程计数器。如果该计数器大于零,就挂起线程。只有当计数器为零时(即,没有线程在等待),才能释放所有等待的任务,然后才能继续执行。

Barrier 对象提供 ParticipantCount 属性和 ParticipantRemaining 属性(前者返回待同步任务的数量,后者返回在到达屏障[*barrier*]之前要调用 SignalAndAwait 的任务数量),能让被挂起的线程继续执行。可以提供 postPhaseAction 方法委托作为 Barrier 构造函数的一个参数,当所有任务到达屏障时将调用这个构造函数。Barrier 对象被作为参数传入 postPhaseAction 方法。在该方法返回之前,Barrier 不可用,任务不能恢复执行。

下面的示例将计算一个数学公式,用 Barrier 对象按顺序绘制图形演示同步。

## 准备就绪

确定安装并运行了 Visual Studio。

## 操作步骤

现在,根据下面的步骤创建程序,稍后再详细解释。

1. 创建一个新的 C++ CLR 空项目,并命名为 CLRBarrier。

2. 打开【解决方案资源管理器】,右键单击【源文件】,添加一个新的源文件 main。打开 main.cpp。输入下面的代码:

```
#include <Windows.h>
#include <tchar.h>
#include "MyForm.h"

using namespace System::Windows::Forms;
using namespace System::Threading;
using namespace CLRBarrier;

[STAThreadAttribute]
int APIENTRY _tWinMain(HINSTANCE hThis, HINSTANCE hPrev, LPTSTR
    szCommandLine, int iCmdShow)
{
    Application::EnableVisualStyles();
    Application::SetCompatibleTextRenderingDefault(false);

    Application::Run(gcnew MyForm());
```

```
        return 0;
}
```

3. 打开【解决方案资源管理器】，右键单击【头文件】。在【Visual C++】下选择 UI，添加一个新的 Windows 窗体，并命名为 MyForm。打开 MyForm.h。

4. 输入下面的代码：

```
#pragma once

namespace CLRBarrier
{
    using namespace System;
    using namespace System::ComponentModel;
    using namespace System::Collections::Generic;
    using namespace System::Windows::Forms;
    using namespace System::Threading;
    using namespace System::Threading::Tasks;
    using namespace System::Drawing;

    public ref class MyForm : public Form
    {
    public:
        MyForm(void)
        {
            InitializeComponent();

            iWidth = 800;
            iHeight = 600;

            bitmapGraph = gcnew Bitmap(iWidth, iHeight);
            colorArray = gcnew array<Color>(iWidth * iHeight);

            color = Color::FromArgb(0, 0, 255);

            progress->Maximum = iWidth / 2;
        }
    protected:
        ~MyForm()
        {
            if (components)
            {
                delete components;
            }
        }
    private:
        System::Windows::Forms::GroupBox^ groupBox;
        System::Windows::Forms::Label^ status;
        System::Windows::Forms::ProgressBar^ progress;
        System::Windows::Forms::Button^ btnStart;
        System::Windows::Forms::PictureBox^ canvas;
        System::ComponentModel::Container ^components;

        int iWidth;
        int iHeight;
        array<Color>^ colorArray;
        Bitmap^ bitmapGraph;
        Barrier^ barrier;
```

```cpp
Color color;

delegate void UpdateProgressDelegate(void);
System::Void UpdateProgress(void)
{
    progress->PerformStep();
}

delegate void SetCanvasDelegate(void);
void SetCanvas(void)
{
    canvas->BackColor = Color::Black;
    canvas->Image = bitmapGraph;
}

delegate void SetLabelTextDelegate(String^ szText);
void SetLabelText(String^ szText)
{
    status->Text = szText;
}

System::Void EndTask(Barrier^ b)
{
    FillBitmap();

    SetLabelTextDelegate^ setLabelText = gcnew
        SetLabelTextDelegate(this, &MyForm::SetLabelText);
    status->BeginInvoke(setLabelText, L"Done.");

    SetCanvasDelegate^ canvasDelegate = gcnew
        SetCanvasDelegate(this, &MyForm::SetCanvas);
    canvas->BeginInvoke(canvasDelegate);
}

System::Void btnStart_Click(Object^ sender, EventArgs^ e)
{
    status->Text = L"Calculating...";

    Action^ action = gcnew Action(this,
        &MyForm::CalculateValues);

    barrier = gcnew Barrier(2, gcnew Action<Barrier^>(this,
        &MyForm::EndTask));

    Tasks::Parallel::Invoke(action, action);
}

System::Void FillBitmap(void)
{
    for (int iYPosition = 0; iYPosition < iHeight; iYPosition++)
    {
        for (int iXPosition = 0; iXPosition < iWidth; iXPosition++)
        {
            bitmapGraph->SetPixel(iXPosition, iYPosition,
                colorArray[(iXPosition + (iWidth * iYPosition))]);
        }
    }
}
```

```cpp
System::Void CalculateValues(void)
{
    int iLeft = iWidth / 2;
    int iArea = iLeft * iLeft;
    int iTop = iHeight / 2;

    UpdateProgressDelegate^ updateProgress = gcnew
        UpdateProgressDelegate(this, &MyForm::UpdateProgress);

    for (int iXAxis = 0; iXAxis < iLeft; iXAxis++)
    {
        int iSurface = iXAxis * iXAxis;
        double dSquare = Math::Sqrt(iArea - iSurface);

        for (double dIndex = -dSquare; dIndex < dSquare; dIndex += 3)
        {
            double dRadius = Math::Sqrt(iSurface + dIndex * dIndex)
                / iLeft;
            double dBottom = (dRadius - .96) * Math::Sin(32 *
                dRadius);
            double dYAxis = dIndex / 3 + (dBottom * iTop);

            SetPixelValue(colorArray, (int)(-iXAxis + (iWidth /
                2)), (int)(dYAxis + (iHeight / 2)));
            SetPixelValue(colorArray, (int)(iXAxis + (iWidth / 2)),
                (int)(dYAxis + (iHeight / 2)));
        }

        progress->BeginInvoke(updateProgress);
    }

    barrier->SignalAndWait();
}

System::Void SetPixelValue(array<Color>^ graphArray, int
    iXPos, int iYPos)
{
    int iIndex = (iXPos + iYPos * iWidth);
    graphArray[iIndex] = color;
}

#pragma region Windows Form Designer generated code
    /// <summary>
    /// 设计者所需的方法,
    /// 不要用代码编辑器修改该方法的内容
    /// </summary>
    void InitializeComponent(void)
    {
        this->groupBox = (gcnew
            System::Windows::Forms::GroupBox());
        this->status = (gcnew
            System::Windows::Forms::Label());
        this->progress = (gcnew
            System::Windows::Forms::ProgressBar());
        this->btnStart = (gcnew
            System::Windows::Forms::Button());
        this->canvas = (gcnew
            System::Windows::Forms::PictureBox());
        this->groupBox->SuspendLayout();
```

```cpp
    (cli::safe_cast
        <System::ComponentModel::ISupportInitialize^>
        (this->canvas))->BeginInit();
this->SuspendLayout();
//
// groupBox
//
this->groupBox->Controls->Add(this->status);
this->groupBox->Controls->Add(this->progress);
this->groupBox->Controls->Add(this->btnStart);
this->groupBox->Location =
    System::Drawing::Point(9, 615);
this->groupBox->Name = L"groupBox";
this->groupBox->Size =
    System::Drawing::Size(800, 59);
this->groupBox->TabIndex = 0;
this->groupBox->TabStop = false;
//
// status
//
this->status->Location =
    System::Drawing::Point(13, 20);
this->status->Name = L"status";
this->status->Size =
    System::Drawing::Size(270, 27);
this->status->TabIndex = 2;
this->status->TextAlign =
    System::Drawing::ContentAlignment::MiddleLeft;
//
// progress
//
this->progress->Location =
    System::Drawing::Point(301, 20);
this->progress->Name = L"progress";
this->progress->Size =
    System::Drawing::Size(387, 27);
this->progress->Step = 1;
this->progress->TabIndex = 1;
//
// btnStart
//
this->btnStart->Location =
    System::Drawing::Point(704, 20);
this->btnStart->Name = L"btnStart";
this->btnStart->Size =
    System::Drawing::Size(81, 27);
this->btnStart->TabIndex = 0;
this->btnStart->Text = L"Start";
this->btnStart->UseVisualStyleBackColor = true;
this->btnStart->Click += gcnew
    System::EventHandler(this,
        &MyForm::btnStart_Click);
//
// canvas
//
this->canvas->Location =
    System::Drawing::Point(9, 9);
this->canvas->Name = L"canvas";
this->canvas->Size =
```

```
                    System::Drawing::Size(800, 600);
                this->canvas->TabIndex = 1;
                this->canvas->TabStop = false;
                //
                // MyForm
                //
                this->AutoScaleDimensions =
                    System::Drawing::SizeF(6, 13);
                this->AutoScaleMode =
                    System::Windows::Forms::AutoScaleMode::Font;
                this->ClientSize =
                    System::Drawing::Size(817, 684);
                this->Controls->Add(this->canvas);
                this->Controls->Add(this->groupBox);
                this->MaximizeBox = false;
                this->MinimizeBox = false;
                this->Name = L"MyForm";
                this->Text = L"The Barrier example";
                this->groupBox->ResumeLayout(false);
                (cli::safe_cast
                    <System::ComponentModel::ISupportInitialize^>
                    (this->canvas))->EndInit();
                this->ResumeLayout(false);
            }
#pragma endregion
    };
}
```

## 示例分析

我们在示例中使用 800×600（像素）的图像，所以把 iWidth 和 iHeight 分别设置为 800 和 600。然后，分配 Bitmap 和颜色数组。由于我们要绘制一个左右对称的图案，所以将用像素宽度的一半进行迭代，而且把进度的最大值设置为为 iWidth / 2。单击 **Start** 按钮时，为了将其传递给 Tasks::Parallel::Invoke 方法，要创建一个方法委托。在调用并行调用程序之前，需要用两个参与者和 postPhaseAction 方法委托实例化屏障对象。对我们而言，要在整个计算完成之后绘制图案，这个寄送行为方法委托非常重要。我们可以用一个事件来完成，或者给窗口消息队列寄送一条消息，这才是使用 Barrier 对象的正确方式。我们的寄送行为方法是 EndTask，它要绘制图案并设置标签文本。由于不同的线程都要调用该方法，而且必须避免用 UI 元素进行跨线程操作，所以我们必须用委托来调用合适的方法。

## 更多讨论

本章介绍的这些.NET 框架的构件都非常好用，但是对于给定的任务，它们都有或多或少的效率问题。.NET 内置了许多供用户在开发过程中使用的构件。程序员要自己探索和深入研究，针对开发过程中遇到的问题，选择那些与任务相匹配的构件。

# 第 7 章 理解并发代码设计

本章介绍以下内容：
- 如何设计并行应用程序
- 理解代码设计中的并行
- 转向并行
- 改进性能因素

## 7.1 介绍

开发并行应用程序时，同步和时间是要重点考虑的因素，如果在程序中添加并发还要考虑诸多其他问题。

并行编程是一把双刃剑。它既可以做很多事情（如，让程序能在最新的多核处理器上执行），也是实现各种软件复杂功能的基础。程序员通过并行编程，能开发出用户界面丰富且快速响应的 GUI 应用，提升用户体验。但是，必须注意并发是否使用得当，否则会导致严重的性能问题。

设计糟糕的并发程序比顺序程序的性能更差，因为 CPU 要在重新调度执行线程、获取内存块、设置同步对象的触发状态、维护保护机制等方面花费巨额开销。

在讨论以并行方式创建应用程序之前，我们先提醒读者注意一下顺序执行（*sequential execution*）。

对于应用程序的生存期，顺序编程非常重要，因为程序的算法决定了应用程序（线程）必须执行的具体步骤。即使在并发应用程序中，选择正确的顺序方式也很重要。我们应该先尽最大努力优化顺序方案，然后再将其并行化。

本章将探讨一些并行编程时可能遇到的麻烦情况及其解决方案，学习程序示例如何正确地使用并行。

## 7.2 如何设计并行应用程序

首先要能识别可能出问题的情形，才能避免在编程时出错。我们把涉及并发的问题分成两类。

- **正确性问题**：导致程序产生错误的结果。
- **活跃性问题**：导致程序停止产生结果。

鉴于并发本身的性质，正确性问题很难被发现。每个线程的执行路径不同，有时会出错。这种情况任何时候都有可能发生，程序员完全意识不到程序输出了错误的结果（因为之前明明一切正常）。最糟糕的情况是，在未出状况之前，同一个程序无论运行多少次都没问题，得到的结果都正常。但是，可能由于使用了不同的参数就会导致程序产生错误的结果。

其中一种可能出问题的情形是**数据竞争**（*data race*）。在所有重要的程序中，都存在一些程序状态和控制流之间关系的重要假设，不能违反这些假设，否则应用程序将生成错误的结果。我们来看下面的例子：

```
int iValue = 5;
...
int iAValue = iValue;
...
int iBValue = iValue;
```

如果有多个线程正在访问共享对象 iValue，为了避免出现意想不到的情况和生成错误的结果，我们必须添加某种安全并发机制，如独占访问或数据同步。

遇到这些情况时，我们必须确定要同步什么内容、应该使用哪些同步对象或技术。

另一种可能出问题的情形是**不一致同步**（*Inconsistent Synchronization*）。有时，通过某种类型的同步机制仅保护访问行为还不够，必须把访问某对象的所有对象（线程）都用相同类型的同步机制保护起来才行。下面的代码中，注意 funcA 和 funcB 方法中用不同的同步对象声明了锁：

```
class Clock
{
public:
    CLock( TCHAR* szMutexName );
    ~CLock( );
private:
    HANDLE hMutex;
};

CLock::CLock( TCHAR* szMutexName )
{
    hMutex = CreateMutex( NULL, FALSE, szMutexName );
    WaitForSingleObject( hMutex, INFINITE );
}

CLock::~CLock( )
{
    ReleaseMutex( hMutex );
    CloseHandle( hMutex );
}

...

int iValue = 5;
void funcA( void )
{
    CLock* lock = new CLock( TEXT( "_tmp_lock_01_" ) );
    ++iValue;
```

```
        delete lock;
    }
    void funcB( void )
    {
        CLock* lock = new CLock( TEXT( "_tmp_lock_02_" ));
        ++iValue;
        delete lock;
    }
```

这种情况和前面数据竞争的例子类似，如果在不同的线程中调用方法，就会得到错误的结果，因为它们的保护机制不一致。在所谓理想情况下，单独的线程只会对共享对象执行读操作。但事实并非如此，这样的代码有潜在的危险。

还有一种可能出问题的情形是**复合动作**（*Composite Action*）。我们经常要权衡是用单线程（更多的锁）来处理，还是用并发（无锁）来处理。这取决于要保护哪些语句或哪部分代码。

假设我们要移除简单链表的头节点。当我们提到权衡，而且要移除链表的头字节时，我们应该有两个临界区。首先，询问头节点是否等于零(0)，即 node == NULL；然后，给新节点赋值，即 next = node->next。如果我们保护许多语句，就会损失并行化。我们要理解权衡的意义：保护的语句越少，完成的并发越多。

假设链表只有一个节点，我们将使用独占访问示例中的 CLock 类。代码如下所示：

```
        template class<T>
        class CList<T>
        {
        public:
            CList( ) : next( 0 ), value( 0 ) { }
            T Pop( void );
        private:
            CList* next;
            T* value;
        };

        template class<T>
        T* CList::Pop( void )
        {
            CList* node = next;
            if (node == NULL)
            {
              return 0;
            }

            CLock* lock = new CLock(_T("_tmp_lock_"));
            next = node->next;
            delete lock;

            return node->value;
        }
```

读者是否发现上面的代码会出现一些问题。虽然我们保护了 assignmen 语句周围的临界区，但是必须时刻注意执行流！如果并行执行这部分代码会怎样？这将导致以下情况：线程 A 进入 Pop 方法并执行 node == NULL 语句，获得 FALSE。然后，操作系统突然转而执行另一个线程。现在，线程 B 进入 Pop 方法，并执行 node == NULL 语句，获得 FALSE。接着，操作系统又转而执行线程 A，线程 A 从上次暂停的地方继续执行。它将获得一个锁，并移除链表的头字节。

当操作系统下次执行线程 B 时，就会出现糟糕的情况。线程 B 从上次暂停的地方开始继续执行，获得一个锁，然后访问已经不存在的对象。这绝对会导致应用程序崩溃！如图 7.1 所示。

| T | 线程 A | 线程 B |
|---|---|---|
| 0 | CList* node = next; | |
| 1 | next == NULL; | |
| 2 | | CList* node = next; |
| 3 | | next == NULL; |
| 4 | CLock* lock = new CLock(_T("_tmp_lock_")); | |
| 5 | next = node->next; | |
| 6 | | CLock* lock = new CLock(_T("_tmp_lock_")); |
| 7 | delete lock; | |
| 8 | | next = node->next; //应用程序崩溃！|

图 7.1

第 1 列（T）是 CPU 执行线程所需的时间。第 2 列是线程 A 的语句，第 3 列是线程 B 的语句。即使设计了保护，也没设计对。要解决这个问题，必须保护更多的语句，如下所示：

```
template class<T>
class CList<T>
{
public:
   CList( ) : next( 0 ), value( 0 ) { }
   T Pop( void );
private:
CList* next;
   T* value;
};

template class<T>
T* CList::Pop( void )
{
   CLock* lock = new CLock(_T("_tmp_lock_"));

   CList* node = next;
   if (node == NULL)
   {
      return 0;
   }

   next = node->next;
   delete lock;

   return node->value;
}
```

现在，即使操作系统切换线程也不会发生**正确性问题**了，因为我们确保了进入 Pop 函数后由多个步骤构成的行为不可分割。然而，保护的语句越多意味着顺序执行越多，所以在设计保护时还要注意性能问题。

在设计应用程序的执行流时，应该把容易出问题的地方设计成语句较少的小型重点保护部分，同时还要尽量从用户角度考虑程序的使用情况。这样才能确定正确地并行执行流。

接下来，我们讨论**活跃性问题**。通常，活跃性问题不会终止程序，因此检测和诊断都非常困难。在大多数情况下，这些问题会导致应用程序停止响应。有时，甚至会导致应用程序暂时停止运行。**死锁**就是其中一种情形。我们来看一个需要在两个银行账户之间转账的例子。我们要实现 CBankAccount 类，可以转账，带有锁保护（万一并行使用账户）。代码如下：

```
class CBankAccount
{
public:
    CBankAccount(){ _tcscpy( szLock, LockName( ) ); }
    static void Transfer(CBankAccount* a, CBankAccount* b, double dAmount);
    static TCHAR* LockName(void);
private:
    unsigned uID;
    double dBalance;
    TCHAR szLock[16];
};

TCHAR* CBankAccount::LockName(void)
{
    static int iCount = 0;
    static TCHAR szBuffer[16];
    wsprintf( szBuffer, _T("_lock_%d_"), ++iCount );
    return szBuffer;
}

void CBankAccount::Transfer(CBankAccount* a, CBankAccount* b, double dAmount )
{
    CLock* outerLock = new CLock(a->szLock);
    if (dAmount < a->dBalance)
    {
        CLock* innerLock = new CLock(b->szLock);
        a->dBalance -= dAmount;
        b->dBalance += dAmount;
        delete innerLock;
    }
    delete outerLock;
}
```

上面的代码看上去没问题，其实并不正确！假设线程 A 要从账户 X 转账到账户 Y，同时，线程 B 要从账户 Y 转账到账户 X。当线程 A 进入 `Transfer` 方法时，将获得属于帐号 X 的锁。现在，假设帐号 X 使用的互斥量名是 `lock_1`。然后，假设系统下一次调度线程 B，它将获得属于帐号 Y 的锁，其互斥量名是 `lock_2`。接下来，线程 A 继续执行，尝试获得属于帐号 Y 的锁。此时，线程 B 正持有属于帐号 Y 的 `lock_2`，于是线程 A 被阻塞了。现在，线程 B 继续执行，尝试获得 `lock_1`，但是 `lock_1` 却被刚才已经被挂起的线程 A 持有。如图 7.2 所示。

| T | 线程 A（从 X 转账到 Y） | 线程 B（从 Y 转账到 X） |
|---|---|---|
| 0 | CLock* outerLock = new CLock(a->LockName());<br>// 线程 A 在帐号 X 中获得 lock_1 | |
| 1 | | CLock* outerLock = new CLock(a->LockName());<br>// 线程 B 在帐号 Y 中获得 lock_2 |
| 2 | if(dAmount < a->dBalance) | |
| 3 | | if(dAmount < a->dBalance) |
| 4 | CLock* outerLock = new CLock(b->LockName());<br>// 线程 A 尝试在帐号 Y 中获得 lock_2，被挂起 | |
| 5 | 被挂起 | CLock* outerLock = new CLock(b->LockName());<br>// 线程 A 尝试在帐号 Y 中获得 lock_2，被挂起 |
| 6 | // 线程永远等待!! | 被挂起 |
| 7 | | // 线程永远等待!! |

图 7.2

第 1 列（T）是执行语句的时间，第 2 列是线程 A 的语句代码，第 3 列是线程 B 的语句代码。图 7.2 演示的这种情况极少发生，但是在设计并行代码时，必须时刻意识到会发生这些情况。

## 7.3 理解代码设计中的并行

我们在书中多次提到，开发过程中最困难的是如何正确地设计应用程序，而不是编写代码。完成程序的设计阶段后，根据设计思路一步一步编写代码并不难。接下来，我们将用同一个示例演示如何从零开始编写并行应用程序。刚开始是最基础的版本，然后我们将逐步完善。当然，在开始设计之初就要考虑到今后会添加并行。

我们将创建一个程序计算 Schwefel's 函数。通常，概率和数理统计用 Schwefel's 函数和类似函数得到的计算结果与采用遗传学算法得到的计算结果相差很大。因为遗传学算法将生成一个巨大的人口基因组（携带各自的特征），然后比较它们的得分（结果）。为了简化问题且满足所需的要求，我们将创建 CSchwefel 类。

Schwefel's 函数如下所示：

$$f(x) = \sum_{i=1}^{n}\left(\sum_{j=1}^{i} x_j\right)^2, -65.536 \le x_i \le 65.536$$

### 准备就绪

确定安装并运行了 **Visual Studio**。

## 操作步骤

现在，根据下面的步骤创建程序，稍后再详细解释。

1. 创建一个新的默认 C++ 控制台应用程序，并命名为 ConcurrentDesign。

2. 打开【解决方案资源管理器】，打开 stdafx.h 文件。输入下面的代码：

    ```
    #pragma once

    #include "targetver.h"

    #include <stdio.h>
    #include <tchar.h>
    #include <Windows.h>
    #include <cfloat>
    #include <time.h>
    #include <omp.h>

    #include "Schwefel.h"
    ```

3. 打开【解决方案资源管理器】，右键单击【头文件】。添加一个新的头文件 Schwefel。打开 Schwefel.h，输入下面的代码：

    ```
    #ifndef __SCHWEFEL__
    #define __SCHWEFEL__

    #include "stdafx.h"

    SYSTEMTIME operator - (const SYSTEMTIME& stFirst, const
        SYSTEMTIME& stSecond);

    class CSchwefel
    {
    public:
        CSchwefel(void) : pdAllele(0) {}
        CSchwefel(unsigned uSize) {
            pdAllele = new double[this->uSize = uSize]; Initialize();
        }
        double* Allele(void) const { return pdAllele; }
        double Allele(unsigned uIndex) const { return pdAllele[uIndex]; }
        double Allele(double dValue, unsigned uIndex) {
            return pdAllele[uIndex] = dValue;   }
        unsigned Size(void) const { return uSize; }
        unsigned Size(unsigned uSize) { return this->uSize = uSize; }
        virtual ~CSchwefel(void) { if (pdAllele != 0) delete[] pdAllele;  }
        double& Sum(void) { return dSum; }
        double& Min(void) { return dMin; }
        double& Score(void) { return dScore; }
        double& Best(void) { return dBest; }
        virtual double Evaluate(void* lpParameters = 0);
        void Initialize();
        void Initialize(unsigned uSize);
        SYSTEMTIME& StartTime(void) { return stStart; }
        SYSTEMTIME& EndTime(void) { return stEnd; }
        TCHAR* Elapsed(void);
        TCHAR* Elapsed(TCHAR* szElapsed);
    ```

```cpp
        static const unsigned ThreadNumber = 4;
    protected:
        CSchwefel(const CSchwefel&) { }
        CSchwefel& operator = (const CSchwefel&) { return *this; }
        virtual TCHAR* ObjectName(void) const { return _T("Schwefel"); }
    private:
        double* pdAllele;
        unsigned uSize;
        double dSum;
        double dMin;
        double dScore;
        double dBest;
        SYSTEMTIME stStart;
        SYSTEMTIME stEnd;
        double Random(double dMin, double dMax);
        static unsigned Seed(void) {
            static unsigned uSeed = rand() % time(NULL);
            return uSeed++;
        }
    };

    #endif
```

4. 打开【解决方案资源管理器】,右键单击【源文件】。添加一个新的源文件 Schwefel。打开 Schwefel.cpp,输入下面的代码:

```cpp
#include "stdafx.h"

double CSchwefel::Evaluate(void* lpParameters)
{
    GetSystemTime(&stStart);

    for (unsigned i = 0; i < uSize; i++)
    {
        dSum += pdAllele[i];
        if (dMin > pdAllele[i])
        {
            dMin = pdAllele[i];
        }

        double dInnerSum = 0;
        for (unsigned j = 0; j < i; j++)
        {
            dInnerSum += pdAllele[j];
        }

        double dTmp = dInnerSum * dInnerSum;
        dScore += dTmp;

        if (i != 0 && dBest > dTmp)
        {
            dBest = dTmp;
        }
    }

    GetSystemTime(&stEnd);

    return dScore;
```

```cpp
}

void CSchwefel::Initialize(void)
{
    srand(Seed());

    dSum = 0;
    dMin = DBL_MAX;
    dScore = 0;
    dBest = DBL_MAX;
    stStart = { 0 };
    stEnd = { 0 };

    for (unsigned uIndex = 0; uIndex < uSize; uIndex++)
    {
        pdAllele[uIndex] = Random(-65.536, 65.536);

        dSum += pdAllele[uIndex];

        if (dMin > pdAllele[uIndex])
        {
            dMin = pdAllele[uIndex];
        }
    }
}

void CSchwefel::Initialize(unsigned uSize)
{
    if (pdAllele == NULL)
    {
        pdAllele = new double[this->uSize = uSize];
    }

    srand(Seed());

    dSum = 0;
    dMin = DBL_MAX;
    dScore = 0;
    dBest = DBL_MAX;
    stStart = { 0 };
    stEnd = { 0 };

    for (unsigned uIndex = 0; uIndex < uSize; uIndex++)
    {
        pdAllele[uIndex] = Random(-65.536, 65.536);

        dSum += pdAllele[uIndex];

        if (dMin > pdAllele[uIndex])
        {
            dMin = pdAllele[uIndex];
        }
    }
}

TCHAR* CSchwefel::Elapsed(void)
{
    static TCHAR szBuffer[1024];
```

```cpp
        SYSTEMTIME stTime = stEnd - stStart;

        swprintf(szBuffer, 1024, _T("%ws:\nAllele size:\t%u\nSum:\t\
t%lf\nMin:\t\t%lf\nScore:\t\t%lf\nBest:\t\t%lf\nExecution time:% 02u: % 02u :
% 02u : % 03u\n\n"),ObjectName(), uSize, dSum, dMin, dScore, dBest, stTime.wHour,
stTime.wMinute, stTime.wSecond, stTime.wMilliseconds);

        return szBuffer;
}

TCHAR* CSchwefel::Elapsed(TCHAR* szElapsed)
{
        if (szElapsed == NULL || _tcslen(szElapsed) < 1024)
        {
                return _T("Error!\nBuffer too small.");
        }

        SYSTEMTIME stTime = stEnd - stStart;

        swprintf(szElapsed, 1024, _T("%ws:\nAllele size:\t%u\nSum:\t\
t%lf\nMin:\t\t%lf\nScore:\t\t%lf\nBest:\t\t%lf\nExecution time:% 02u : % 02u : %
 02u : % 03u\n\n"),ObjectName(), uSize, dSum, dMin, dScore, dBest, stTime.wHour,
stTime.wMinute, stTime.wSecond, stTime.wMilliseconds);

        return szElapsed;
}

double CSchwefel::Random(double dMin, double dMax)
{
        double dValue = (double)rand() / RAND_MAX;
        return dMin + dValue * (dMax - dMin);
}

SYSTEMTIME operator - (const SYSTEMTIME& stFirst, const
        SYSTEMTIME& stSecond)
{
        SYSTEMTIME stResult;
        FILETIME ftTime;
        ULARGE_INTEGER uLargeInteger;

        __int64 i64Right;
        __int64 i64Left;
        __int64 i64Result;

        SystemTimeToFileTime(&stFirst, &ftTime);
        uLargeInteger.LowPart = ftTime.dwLowDateTime;
        uLargeInteger.HighPart = ftTime.dwHighDateTime;
        i64Right = uLargeInteger.QuadPart;

        SystemTimeToFileTime(&stSecond, &ftTime);
        uLargeInteger.LowPart = ftTime.dwLowDateTime;
        uLargeInteger.HighPart = ftTime.dwHighDateTime;
        i64Left = uLargeInteger.QuadPart;

        i64Result = i64Right - i64Left;

        uLargeInteger.QuadPart = i64Result;
        ftTime.dwLowDateTime = uLargeInteger.LowPart;
        ftTime.dwHighDateTime = uLargeInteger.HighPart;
```

```
            FileTimeToSystemTime(&ftTime, &stResult);

            return stResult;
        }
```

5. 打开【解决方案资源管理器】，打开 ConcurrentDesign.cpp。输入下面的代码：

```
    #include "stdafx.h"

    int _tmain(int argc, _TCHAR* argv[])
    {
        CSchwefel Schwefel(100000);
        Schwefel.Evaluate();

        _tprintf_s(Schwefel.Elapsed());

    #ifdef _DEBUG
        return system("pause");
    #endif

        return 0;
    }
```

## 示例分析

首先，我们定义了 CSchwefel 类。该类用于储存 double 类型的数组，其元素的取值范围是 −65.536~65.536，这同时也是 Schwefel's 方程的取值范围。该数组名为 allele，与基因组结构类似。另外，我们还需要知道 allele 数组的大小、元素的个数、最小值、分数、最佳值以及开始和结束的时间。

我们有两个自带 allele 大小的默认构造函数。如果使用默认构造函数，创建的是 "假的" 对象，在计算之前必须调用 Initialize 方法进行初始化。当然，也可以使用内置的初始化构造函数，但是必须提供 allele 的大小。对于我们的示例，第 2 种构造函数显然更好。除此之外，还要为类的特征实现 set 和 get 函数。double* Allele(void) 返回指向 allele 数组的指针；double Allele(unsigned uIndex) 返回 uIndex 指定数组索引上的值；double Allele(double dValue, unsigned uIndex) 把 uIndex 的值设置为 dValue，并返回新添加的值。unsigned Size(void) 返回 allele 数组的大小，unsigned Size(unsigned uSize) 设置新的数组大小，并返回新添加的大小。

Sum、Min、Score、Best、StartTime 和 EndTime 方法分别返回类特征 dSum、dMin、dScore、dBest、stStart 和 stEnd 的引用，因此我们也可以把这些方法作为 get 和 set 方法来用。Initialize(void) 和 Initialize(unsigned uSize) 方法用 −65536~65536 之间的随机值填充 allele 数组，并同时计算 allele 数组的总和和最小值。

Evaluate(void*) 方法和析构函数一样声明为虚函数，因为派生类可以改变求值顺序或稍后添加并发。Elapsed(void) 和 Elapsed(TCHAR*) 方法把输出写入控制台。我们还添加了常量值 ThreadNumber，用于设置允许的最大线程。可以任意改变该值。

我们把拷贝构造函数和 `operator=` 的重载方法放在 `protected` 域中，以防止程序员使用它们。因为默认拷贝操作会把类特征的值逐一拷贝，如果拷贝的都是整型值不会出什么问题，但是如果有指针就麻烦了。拷贝指针的值还不够，这意味着两个对象的指针都指向相同的数组或指针指向的任意内存。必须实现拷贝构造函数和=操作符，为类对象的每一个实例分配新的内存，然后再逐一拷贝旧指针指向的值。`TCHAR* ObjectName(void)` 方法在格式化 `Elapsed` 方法输出的字符串时返回类名。在派生类中，我们将把 `ObjectName` 方法重载为返回正确的重载类名。

在 `private` 域中，有 `Random(double dMin, double dMax)` 和 `Seed(void)` 方法，分别返回 `dMin` 到 `dMax` 之间的随机 `double` 值和随机生成器的种子。

前面提到过，`Evaluate` 方法被声明为 `virtual`，对于 `CSchwefel` 类，我们直接实现就可以，暂时不必考虑其他情况。首先，为了在开始计算之前储存时间，以 `stStart` 作为参数调用 `GetSystemTime` API。接着，在外层循环中计算总和和最小值，在内层循环中计算内层总和，然后把结果相乘。根据 Schwefel's 方程，把乘积的结果加给得分（`dScore`），然后计算最佳值。外层循环结束时，为了获得再次获得系统时间，我们以 `stEnd` 为参数调用 `GetSystemTime`。把两个时间相减，得到计算方程花费的时间。另外，还要注意 `Evaluate` 方法。为了方便今后扩展（例如，类派生），我们把 `Evaluate` 方法的参数类型设计为 `void*`。

由于 `SYSTEMTIME` 结构没有为 `SYSTEMTIME` 实现减法，我们只能自己设计：`SYSTEMTIME operator - (const SYSTEMTIME& stFirst, const SYSTEMTIME& stSecond);`。在该方法中，我们使用 `SystemTimeToFileTime` API 把系统时间转换为文件时间，也就是说把符合我们阅读习惯的时间转换为整数值。然后，把两个整数相减，并用 `FileTimeToSystemTime` API 将结果转换为符合我们阅读习惯的时间。

在应用程序的入口，我们用大小为 100000 的 `allele` 数组实例化 `CSchwefel` 对象，并调用 `Evaluate` 和 `Elapsed` 方法。

### 更多讨论

当我们谈论某种技术时，这种技术都不会是解决问题的唯一途径。通常，无论你怎么实现，总还可以用其他方法或技术来完成。我们的初衷是教会读者如何正确地思考和解决与并发相关的问题。要想熟悉并掌握所学的知识，还是要多动手操作，实践是最好的老师。纸上得来终觉浅，绝知此事要躬行。

## 7.4 转向并行

现在，我们要在该示例中添加并行。注意，我们在设计 `CSchwefel` 类时已经把这个考虑进去了。除了 `Evaluate` 方法，其他的内容都不用动。这里，面向对象编程发挥了很大作用，我们不用编写太多的代码，这节约了很多时间。

### 准备就绪

确定安装并运行了 Visual Studio。

### 操作步骤

现在，根据下面的步骤创建 CSchwefelMT 类，稍后再详细解释。

1. 打开【解决方案资源管理器】，打开 stdafx.h 文件。输入下面的代码：

    ```
    #pragma once

    #include "targetver.h"

    #include <stdio.h>
    #include <tchar.h>
    #include <Windows.h>
    #include <cfloat>
    #include <time.h>
    #include <omp.h>

    #include "Schwefel.h"
    #include "SchwefelMT.h"
    ```

2. 打开【解决方案资源管理器】，右键单击【头文件】。添加一个新的头文件 SchwefelWT。打开 SchwefelWT.h，输入下面的代码：

    ```
    #ifndef __SCHWEFELMT__
    #define __SCHWEFELMT__

    #include "stdafx.h"

    class CSchwefelMT : public CSchwefel
    {
    public:
        CSchwefelMT() : CSchwefel() { }
        CSchwefelMT(unsigned uSize) : CSchwefel(uSize) { }
        virtual ~CSchwefelMT(void) { }
        virtual double Evaluate(void* lpParameters = 0);
    protected:
        CSchwefelMT(const CSchwefel&) { }
        CSchwefelMT& operator = (const CSchwefel&) { return *this; }
        virtual TCHAR* ObjectName(void) const {
            return _T("SchwefelMT");
        }
    private:
        static DWORD WINAPI StartAddress(LPVOID lpParameter);
        struct ThreadParams
        {
            CSchwefelMT* this_ptr;
            unsigned uThreadIndex;
        };
    };
    ```

```
#endif
```

3. 打开【解决方案资源管理器】，右键单击【源文件】。添加一个新的源文件 SchwefelWT。打开 SchwefelWT.cpp，输入下面的代码：

```
#include "stdafx.h"
DWORD WINAPI CSchwefelMT::StartAddress(LPVOID lpParameter)
{
    ThreadParams* params = (ThreadParams*)lpParameter;

    unsigned uChunk = ((unsigned)params->this_ptr->Size()) / ThreadNumber;
    unsigned uMax = (params->uThreadIndex + 1) * uChunk;

    for (unsigned i = params->uThreadIndex * uChunk; i < uMax; i++)
    {
        params->this_ptr->Sum() += params->this_ptr->Allele(i);

        if (params->this_ptr->Min() > params->this_ptr->Allele(i))
        {
            params->this_ptr->Min() = params->this_ptr->Allele(i);
        }

        double dInnerSum = 0;

        for (unsigned j = 0; j < i; j++)
        {
            dInnerSum += params->this_ptr->Allele(j);
        }

        double dTmp = dInnerSum * dInnerSum;
        params->this_ptr->Score() += dTmp;

        if (i != 0 && params->this_ptr->Best() > dTmp)
        {
            params->this_ptr->Best() = dTmp;
        }
    }

    return 0L;
}

double CSchwefelMT::Evaluate(void* lpParameters)
{
    HANDLE hThreads[ThreadNumber];
    ThreadParams params[ThreadNumber];

    GetSystemTime(&StartTime);

    for (unsigned uIndex = 0; uIndex < ThreadNumber; uIndex++)
    {
        params[uIndex] = { this, uIndex };
        hThreads[uIndex] = CreateThread(NULL, 0,
            (LPTHREAD_START_ROUTINE)StartAddress, &params[uIndex], 0, NULL);
    }

    WaitForMultipleObjects(ThreadNumber, hThreads, TRUE, INFINITE);
```

```
            GetSystemTime(&EndTime());

            for (unsigned uIndex = 0; uIndex < ThreadNumber; uIndex++)
            {
                CloseHandle(hThreads[uIndex]);
            }

            return Score();
        }
```

4. 打开【解决方案资源管理器】,打开 ConcurrentDesign.cpp 文件。输入下面的代码:

```
        #include "stdafx.h"

        int _tmain(int argc, _TCHAR* argv[])
        {
            CSchwefel Schwefel(100000);
            Schwefel.Evaluate();

            _tprintf_s(Schwefel.Elapsed());

            CSchwefelMT SchwefelMT(100000);
            SchwefelMT.Evaluate();

            _tprintf_s(SchwefelMT.Elapsed());
        #ifdef _DEBUG
            return system("pause");
        #endif

            return 0;
        }
```

## 示例分析

CSchwefelMT 是 CSchwefel 类的派生类,我们只需要实现构造函数、析构函数和 Evaluate 虚函数。和 CSchwefel 类一样,也不允许程序员使用拷贝构造函数和赋值操作符。在 private 域中,我们为线程添加了 StartAddress 例程,为了给线程传递参数添加了 ThreadParams 结构。

Evaluate 方法现在怎么实现?非常简单。我们将实例化线程的 ThreadNumber 和线程参数的 ThreadNumber。然后,调用 GetSystemTime 并运行已声明的线程。我们要创建一个屏障,也就是说,要用 WaitForMultipleObjects 例程等待线程都返回,然后再次调用 GetSystemTime,并关闭所有的线程句柄。

StartAddress 是否可以直接实现而不考虑其他情况?我们来看看。如果按照 7.3 节中的单线程示例逻辑,本例的 StartAddress 方法就有很多正确性问题!我们来解释一下。假设有两个线程(多于两个线程的情况一样),注意下面这行:

```
        params->this_ptr->Sum() += params->this_ptr->Allele(i);
```

在两个线程中访问 dSum 特征完全没有任何保护！如图 7.3 所示。

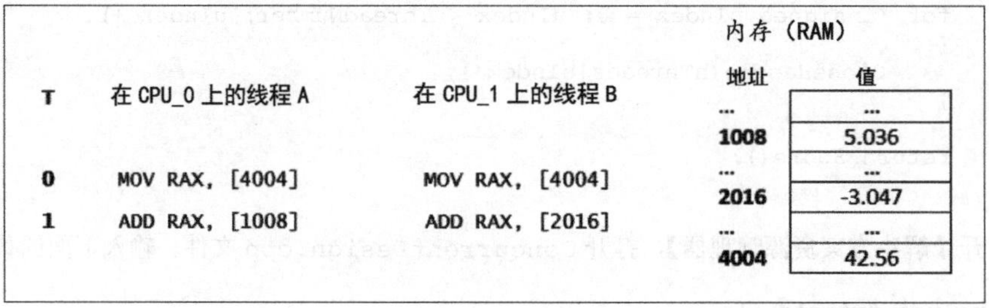

| T | 在 CPU_0 上的线程 A | 在 CPU_1 上的线程 B | 内存（RAM） | |
|---|---|---|---|---|
| | | | 地址 | 值 |
| | | | ... | ... |
| | | | 1008 | 5.036 |
| 0 | MOV RAX, [4004] | MOV RAX, [4004] | ... | ... |
| 1 | ADD RAX, [1008] | ADD RAX, [2016] | 2016 | -3.047 |
| | | | ... | ... |
| | | | 4004 | 42.56 |

**图 7.3**

线程 A 要把 Allele(i) 的值添加给 dSum。Allele(i) 和 dSum 都是有具体地址的内存区域。假设 dSum 的地址是 4004，当前的值是 42.56；假设线程 A 的 Allele(i) 的地址是 1008；线程 B 也要把自己的 Allele(i) 值添加给当前的 dSum 值。dSum 的地址相同（4004），但是线程 A 和线程 B 的 Allele(i) 地址不同（假设线程 B 的 Allele(i) 的地址是 2016）。线程 A 在 CPU_0 上执行，线程 B 在 CPU_1 上执行。

如图 7.3 所示，在时间 0 时，线程 A 的语句是 RAX = 42.56，线程 B 的语句是 RAX = 42.56。在时间 1 时，线程 A 的语句是 RAX = 42.56 + 5.036（结果是 47.596），线程 B 的语句是 RAX = 42.56 + (-3.047)（结果是 39.513）。这两个结果都完全不对！dSum 的值应该是原值加上两个线程的 Allele(i) 值才对（即应该是 44.549）。再次强调，正确设计应用程序是实现并发的关键！

在符合 Schwefel's 函数的计算逻辑的前提下，还要考虑多线程的情况，因此在某种程度上我们改变了 Evaluate 方法。否则，肯定会碰到正确性问题或活跃性问题。因此，StartAddress 方法的正确实现应该如下所示：

```
DWORD WINAPI CSchwefelMT::StartAddress(LPVOID lpParameter)
{
    ThreadParams* params = (ThreadParams*)lpParameter;

    unsigned uChunk = ((unsigned)params->this_ptr->Size()) / ThreadNumber;
    unsigned uMax = (params->uThreadIndex + 1) * uChunk;

    double dSum = 0;
    double dMin = DBL_MAX;
    double dScore = 0;
    double dBest = DBL_MAX;

    HANDLE hMutex = CreateMutex(NULL, FALSE, _T("__tmp_mutex__"));
    WaitForSingleObject(hMutex, INFINITE);

    dSum = params->this_ptr->Sum();
    dMin = params->this_ptr->Min();
    dScore = params->this_ptr->Score();
    dBest = params->this_ptr->Best();

    ReleaseMutex(hMutex);
```

```cpp
        CloseHandle(hMutex);

        for (unsigned i = params->uThreadIndex * uChunk; i < uMax; i++)
        {
            dSum += params->this_ptr->Allele(i);

            if (dMin > params->this_ptr->Allele(i))
            {
                dMin = params->this_ptr->Allele(i);
            }

            double dInnerSum = 0;
            for (unsigned j = 0; j < i; j++)
            {
                dInnerSum += params->this_ptr->Allele(j);
            }
            double dTmp = dInnerSum * dInnerSum;
            dScore += dTmp;

            if (i != 0 && dBest > dTmp)
            {
                dBest = dTmp;
            }
        }

        hMutex = CreateMutex(NULL, FALSE, _T("__tmp_mutex__"));
        WaitForSingleObject(hMutex, INFINITE);

        params->this_ptr->Sum() = dSum;
        if (dMin < params->this_ptr->Min())
        {
            params->this_ptr->Min() = dMin;
        }
        params->this_ptr->Score() = dScore;
        if (dBest < params->this_ptr->Best())
        {
            params->this_ptr->Best() = dBest;
        }

        ReleaseMutex(hMutex);
        CloseHandle(hMutex);

        return 0L;
    }
```

如上代码所示，我们重新设计了语句，这样所有对共享内存区域的访问都是受保护的了。移除了循环中共享对象的引用，因为在循环中无法很好地兼顾使用共享对象和保护访问。那样的代码会只会拖慢应用程序，因为把大部分语句放在循环中保护起来就损失了并发计算。这就是我们添加线程局部存储的原因，这样才能执行并行计算。`StartAddress` 方法的开始部分，我们必须保护那些要读取共享对象的语句，然后使用线程局部变量；在末尾部分再次保护共享对象，然后才执行写操作。这样编写的代码不仅很安全而且逻辑清晰！

## 更多讨论

当我们要改善应用程序的性能时，应该考虑使用并行来完成相同的任务。但是，必须注意两件事：加速性能和效率。如果要对比顺序算法和并行算法的性能，这两件事就很重要。

在现实中，肯定会遇到为了满足重要客户的某些特定要求而微调应用程序的情况。对于这些要求，是否使用并发要慎重决定。如果着重提升性能，就应该创建强大的顺序算法，然后巧妙地将其转为并行。同时，还必须考虑一些副作用，如共享对象或访问违例。

这些情况就像是杠杆平衡（跷跷板）。倾向一边（例如，要改进应用程序的性能）就要损失另一边。我们可以使用并行方法做更多的事。如果使用得当，应用程序会从中受益。另外，从一开始就应该好好考虑使用什么类型的同步对象。

## 7.5 改进性能因素

前面我们提到过，在转向并行之前要构件强大的顺序算法。其中还提到了一些参数，如加速性能和效率，在考虑是否要使用并行时就要参照这些变量。那么，如何测量这些变量？如果执行并行，应用程序是否能从中受益，或者如何大幅提高收益？

本节将尝试回答这些问题。我们将在示例中演示 Visual Studio 编译器的一个内置特性，要依赖 OpenMP，这是英特尔和美国利弗莫尔国家实验室合作开发的 API。OpenMP 代表 Open multiprocessor 编译，是 OpenMP API 的微软实现。它有极优良的内置特性，支持编译器的本地并行执行。

首先，讨论加速。我们如何确定是否使用并发？一个方法是度量和比较执行时间。另一个方法是用某个用于表示以并行方式执行应用程序的数学公式来验证。然而，如果选择第 2 种方法，还是要测试时间。

为了测量这些值，我们必须引入一些变量。如果假设 T(S) 是执行顺序算法所需的时间， T(P) 是执行并行算法所需的时间，那么有下面的等式：

$$I = \frac{T(S)}{T(P)}$$

*I* 代表实现。等式的结果有如下几种情况：

- *I* < 1 表示减速，意味着并行执行只会降低性能；
- *I* < *P* 表示次线性改进；
- *I* ~ *P* 表示线性改进；
- *I* > *P* 表示超线性改进。

减速通常说明用顺序算法来实现性能更好，但也不一定，还可能说明并行代码设计得不好或者进行了太多没必要的保护和同步。还可能是实现错了。理论上，顺序算法本身可以转换成并行算法。但是，要注意另

## 7.5 改进性能因素

一个问题:硬件是否支持并行?在四核 CPU 上测试得出很好的结果,在双核 CPU 上运行会怎样?只考算法本身的话,应用程序的执行时间会比在单核 CPU 上使用顺序算法的执行时间更长。在决定选择什么方式和应用程序应该如何从中受益之前,这些因素都要考虑。其中,下面两个要重点考虑。

- 应用程序如何在更短时间内做更多的事?
- 应用程序如何在不影响系统或其他应用程序的前提下使用更多的资源?

下面的示例将再次计算 Schwefek's 总和。我们将从 CSchwefel 基类派生出一个新类 CSchwefelP,该类将使用 OpenMP 特性来演示另一种能改善程序性能的技术,这样就能时刻关注程序的提升潜力。

为了使用 OpenMP 指令,必须正确设置 Visual Studio 项目的属性页面。具体设置步骤请参见附录。

### 准备就绪

确定安装并运行了 Visual Studio。

### 操作步骤

现在,根据下面的步骤创建程序,稍后再详细解释。

1. 打开解决方案 ConcurrentDesign。

2. 打开【解决方案资源管理器】,右键单击【头文件】。添加一个新的头文件,命名为 SchwefelP。打开 SchwefelP.h。

3. 输入下面的代码:

```
#ifndef __SCHWEFELP__
#define __SCHWEFELP__

#include "stdafx.h"

class CSchwefelP : public CSchwefel
{
public:
    CSchwefelP() : CSchwefel() { }
    CSchwefelP(unsigned uSize) : CSchwefel(uSize) { }
    virtual ~CSchwefelP(void) { }
    virtual double Evaluate(void* lpParameters = 0);
protected:
    CSchwefelP(const CSchwefel&) { }
    CSchwefelP& operator = (const CSchwefel&) { return *this; }
    virtual TCHAR* ObjectName(void) const { return _T("SchwefelP"); }
private:
    //
};

#endif
```

4. 打开【解决方案资源管理器】。右键单击【源文件】。添加一个新的源文件，命名为 SchwefelP。打开 SchwefelP.cpp。

5. 输入下面的代码：

```cpp
#include "stdafx.h"
double CSchwefelP::Evaluate(void* lpParameters)
{
    GetSystemTime(&StartTime());

    double sum = 0;
    double score = 0;
    double min = DBL_MAX;
    double best = DBL_MAX;
    int iSize = (int)Size();

#pragma omp parallel firstprivate(min) firstprivate(best)
    {
#pragma omp for reduction(+: sum) reduction(+: score) nowait
        for (int i = 0; i < iSize; i++)
        {
            sum += Allele(i);

            if (min > Allele(i))
            {
                min = Allele(i);
            }

            double dInnerSum = 0;

            for (int j = 0; j < i; j++)
            {
                dInnerSum += Allele(j);
            }

            double dTmp = dInnerSum * dInnerSum;
            score += dTmp;

            if (i != 0 && best > dTmp)
            {
                best = dTmp;
            }
        }

#pragma omp critical
        {
            if (Best() > best)
            {
                Best() = best;
            }
            if (Min() > min)
            {
                Min() = min;
            }
        }
    }
```

```
        Sum() = sum;
        Score() = score;

        GetSystemTime(&EndTime());

        return Score();
    }
```

6. 打开【解决方案资源管理器】，双击 stdafx.h。

7. 添加下面的代码：

    ```
    #pragma once

    #include "targetver.h"

    #include <stdio.h>
    #include <tchar.h>
    #include <Windows.h>
    #include <cfloat>
    #include <time.h>
    #include <omp.h>

    #include "Schwefel.h"
    #include "SchwefelMT.h"
    #include "SchwefelP.h"
    ```

8. 打开【解决方案资源管理器】，双击 ConcurrentDesign.cpp。

9. 输入下面的代码：

    ```
    #include "stdafx.h"

    int _tmain(int argc, _TCHAR* argv[])
    {
        CSchwefel Schwefel(100000);
        Schwefel.Evaluate();

        _tprintf_s(Schwefel.Elapsed());

        CSchwefelMT SchwefelMT(100000);
        SchwefelMT.Evaluate();

        _tprintf_s(SchwefelMT.Elapsed());

        CSchwefelP SchwefelP(100000);
        SchwefelP.Evaluate();

        _tprintf_s(SchwefelP.Elapsed());

    #ifdef _DEBUG
        return system("pause");
    #endif

        return 0;
    }
    ```

## 示例分析

首先，解释一下派生类 CSchwefelP。我们必须实现构造函数、析构函数和虚函数 Evaluate。其中构造函数和析构函数直接实现就行了，重点在 Evaluate 方法的实现。首先，要获得当前的系统时间。然后，和 CSchwefelMT 类相似，要实例化用于私有和共享对象的线程局部存储对象。注意，我们必须严格遵循并发逻辑。所有要被并行访问的类特征都必须时刻保护！所以，dSum、dScore、dMin 和 dBest 任何时候都不允许有不一致访问。为了只获得 allele 数组大小一次，我们还要添加局部变量 iSize。因为如果把 i < Size() 语句放在并行区内就起不到作用了，而且会带来共享对象的不一致访问问题。

现在，我们使用 OpenMP 指令 #pragma omp parallel，以及子句 firstprivate(min) 和 firstprivate(best)。omp parallel 指令命令编译器定义一个并行构造（并行区），里面的代码将以并行的方式执行多个线程。firstprivate 子句指定了每个线程应该有自己的变量实例，而且变量应该由变量的值初始化，因为该值在构建并行区之前就存在了（摘自 MSDN）。我们的意图是让 min 和 best 的值属于线程私有。因为所有的线程都要计算自己 allele 数组中的最小值和最佳分数。这样，我们才能确保把所有对共享类特征的访问都保护起来了，在本例中是 dMin 和 dBest。

#pragma omp for reduction(+: sum) reduction(+: score) nowait 后面是 for 循环。#pragma omp [parallel] for 指令使得并行区内 for 循环中的工作将被划分成多个线程来执行（摘自 MSDN）。reduction(操作，变量) 子句指定了每个线程私有的一个或多个变量是并行区末尾的归约（*reduction*）操作的对象。这意味着编译器将使用 reduction 子句中指定的变量（每个线程私有），为每个线程进行 reduction 子句中指定的操作，而且在并行区末尾安全地执行必要的归约，本例中是 sum。因此，无论何时执行并行循环，每个线程都将单独计算总和。在 for 循环最后且退出循环之前，编译器将把各线程私有的 sum 值加起来成为一个值。以这种方式，并行就能正确运行，因为所有线程的私有工作都与共享的 sum 区分开来了。每个线程分别计算自己 allele 数组的总和，不会相互影响。在循环末尾，所有线程的 sum 值将被加成一个值。现在，我们要做的是设置类特征。nowait 子句还有一件事：它明确告诉编译器，线程在完成自己循环中的工作后，不必等待其他线程循环完毕。可以把 nowait 看作是一种规定，即在所有线程都达到某点之前，不存在要线程等待的屏障。

在并行循环完毕后，我们添加了另一个并行区：#pragma omp critical。它明确告诉编译器，对该构造下面的语句执行独占访问。每个线程都有自己的最小值和最佳分数。我们必须顺序执行每个线程的比较和赋值，因为同步访问。在根并行构造（*root parallel construct*）的末尾，我们必须为 dSum 和 dScore 赋值，并再次获得系统时间，计算经过了多少时间。

## 更多讨论

我们之前提到测量性能，该示例的输出如图 7.4 所示：

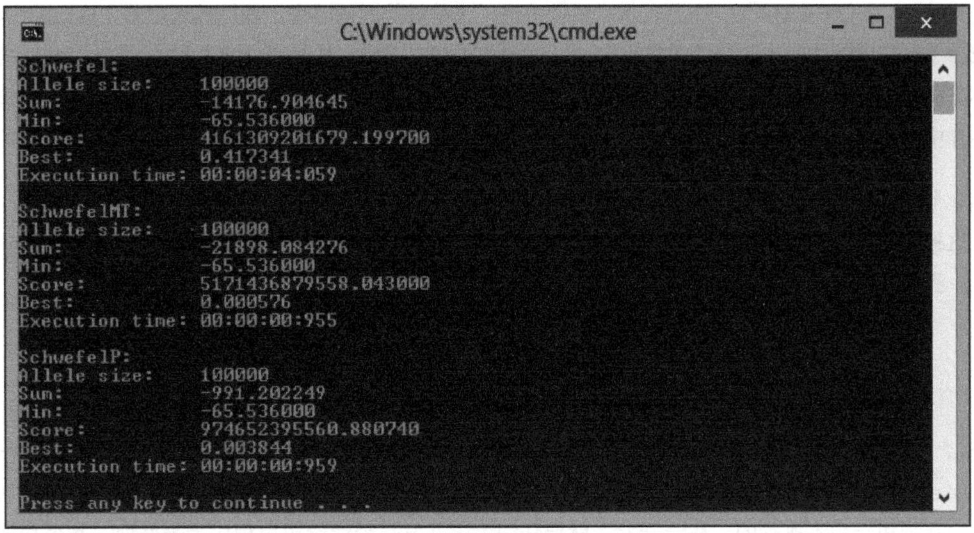

图 7.4

我们使用的是 i7-3820 处理器，有 8 个逻辑核，支持**超线程**（*Hyper-Threading*）技术，可以运行需要硬件支持的 8 个并行线程。

超线程技术能更有效地使用处理器资源，能让多个线程运行在不同的核上。作为性能特性，英特尔的超线程技术还增加了处理器的吞吐量，改进了多线程软件的整体性能。欲详细了解英特尔超线程，请访问下面的链接：

http://www.intel.com/content/www/us/en/architecture-and-technology/hyper-threading/hyper-threading-technology.html。

为了比较 CSchwefelMT 对象和 CSchwefelP 对象的性能，我们把 CSchwefelMT::ThreadNumber 特征设置为 8。这样设置后，运行程序就得出上面的结果。

现在，我们使用改进方程：

$$l = \frac{T(S)}{T(P)}$$

第 1 个输出是 CSchwefel 对象，用顺序算法，单线程。执行时间是 4 秒 57 毫秒。所以 *T(S)* 是 4057（以毫秒为单位）。第 2 个输出是 CSchwefelMT 对象，8 个线程，执行时间是 955 毫秒。第 2 个方法，我们完成了下面的改进：

$$l = \frac{4057}{955} = 4.248$$

结果是 $I < P$（$I = 4.248$，$P = 8$）表示次线性改进，这是最普通的改进。线性改进很难完成，因为程序中或多或少总要使用某些同步和频繁的线程通信。这些会影响完美的线性改进，所以很难完成。而超线性改进几乎不可能完成。如果我们有 X 个处理器，运行某应用程序是否能快 X 倍？在某些情况下，共享对象可以通过储存中间结果加速应用程序，但仅仅是可能。在储存计算了一半的斐波那契数列树时就类似这种情况。

第 3 个输出是 CSchwefelP 对象，8 个线程。我们没有用 `num_threads(int)` 显式设置 `omp parallel` 构造。这种情况类似编译器在创建动态线程池时使用所有可用的线程。它的执行时间是 959 毫秒。对于第 3 种方法，我们完成了下面的改进：

$$l = \frac{4057}{959} = 4.231$$

$I < P$ 说明是次线性改进。相对于 CSchwefelMT 对象而言，这次改进的性能微乎其微，这与许多方面有关，如 `allele` 的大小、OpenMP API 实现、调度器优先级等。我们的意图是给读者演示从设计到编码必须是如何计划和一步步执行的。在一开始，就要仔细考虑并发是否能给应用程序带来好处，是否利大于弊，并使用本书介绍的技术进行改进。实践是最好的老师。成功解决的问题越多，解决问题的能力就越强。

▶ 欲详细了解 OpenMP，请访问：http://openmp.org/wp/ 或者 http://msdn.microsoft.com/en-us/library/tt15eb9t.aspx。

# 第 8 章 高级线程管理

**本章介绍以下内容：**
- 使用线程池
- 定制线程池分发器
- 使用远程线程

## 8.1 介绍

要创建复杂的应用程序，必须包含某些支持复杂设计的智能机制。在我的从业生涯中，见过许多设计良好的项目却选错了软件的解决方案。这些失误包括设计的类不正确，或使用了错误的例程。另外，要特别注意的是，如果在代码中添加了并行，还要加入负责分发和控制并发线程的管理机制。

现在，大部分应用程序和 Windows 操作系统都有一些类似的机制。一种机制是 Windows **分发器对象**（*Windows Dispatcher object*），负责管理和处理线程、调度线程在 CPU 上执行，并维护同步。另一种机制是 Windows 虚拟内存管理器（*Windows Virtual Memory manager*），负责管理内存操作、维护不可访问的内存（这部分内存预留给操作系统或零指针赋值使用）和供应用程序使用的用户空间内存。

这些机制有一个共性，即它们都是根对象的某种抽象（这些根对象负责管理具体的任务）。这对正确维护同步和独占访问非常重要。我们需要一个对象或者管理者来创建和管理任务，以确保所需的资源可用。而且，为了确保只执行一个有效任务或同步操作，这个管理者还必须控制任务的执行。除此之外，还要负责正确销毁任务和释放已使用的资源，以备后用。

其中一种抽象就是线程池，负责并行地执行任务。设计复杂的应用程序或重要的并发操作时，用户肯定会涉及某些类型的线程等级，即某线程比其他线程更重要，以某种方式管理着其他线程。在这种情况下，用户必须维护好已使用的系统资源，以正确的方式加入独占访问、同步等。鉴于这些原因，用户应该先设计一个既能满足上述要求又能服务于开发过程的智能对象。

## 8.2 使用线程池

本节将解释线程池。线程池可表现为工人线程（*worker thread*）的列表，当线程池可用时，将执行用户要求的任务。出现以下两种情形时，使用线程池能让应用程序受益。一种情形是，应用程序要减少可能产生大量线程的数量，并提供工作线程的管理。另一种情形是，为处理小任务创建大量线程。在这种情况中，创建和销毁大量线程所占用的开销巨大，应该用较小数量的线程执行整个工作。

图 8.1 是一个典型的线程池实现。

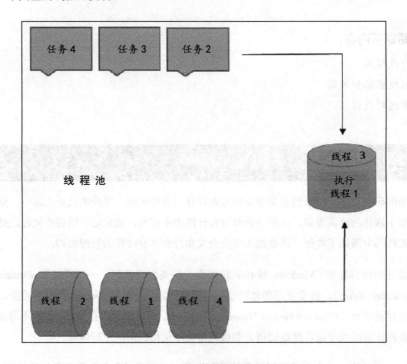

图 8.1　典型的线程池实现

我们将根据这个模式设计 ThreadPool 类。我们将使用一个线程列表，供分发使用。也就是说，应用程序将向线程池请求工人线程，而不是创建线程。线程池的管理者，即分发器，将根据应用程序的要求管理和调度工作线程。

分发器不仅要负责资源的管理、更改和通知，还要负责维护同步和解决问题。应用程序要并行地执行工作，所以要保证足够的系统资源，同时还要注意不能让系统超载太多。另外，为了在某些事件发生时（如，用户已完成输入或后台完成了 I/O 操作）通知工作线程，我们还需要一个警告机制。

因为要并行执行，所以各工作项之间必须同步。我们设计的线程池必须提供诸如线程局部存储这样的特性，供工作项保存/加载成功执行任务所需的内容。另一个重要的特性是能解决潜在的问题。在并行环境中，有些问题是随机情况导致的结果。虽然应用程序看上去设计完美，但是第 7 章中分析的死锁情况也可能发生。因此，必须为线程池小心设计独占访问特性和强大的保护机制。一旦突然发生死锁，分发器必须能处理这种情况，确保任务正常执行。

## 8.2 使用线程池

下面的示例将实现线程池,并使用线程池在两个银行账户之间转移资金。

### 准备就绪

确定安装并运行了 Visual Studio。

### 操作步骤

现在,根据下面的步骤创建程序,稍后再详细解释。

1. 创建一个新的空 C++ 控制台应用程序,并命名为 `ThreadPool`。

2. 打开【解决方案资源管理器】,右键单击【头文件】。添加一个新的头文件 main。打开 main.h,输入下面的代码:

    ```
    #ifndef __MAIN__
    #define __MAIN__

    #include <Windows.h>
    #include <tchar.h>
    #include <time.h>

    #endif
    ```

3. 打开【解决方案资源管理器】,右键单击【头文件】,添加一个新的头文件 CBankAccount。输入下面的代码:

    ```
    #ifndef __BANKACCOUNT__
    #define __BANKACCOUNT__

    #include "main.h"

    class CLock
    {
    public:
        CLock(TCHAR* szMutexName);
        ~CLock();
    private:
        HANDLE hMutex;
    };

    inline CLock::CLock(TCHAR* szMutexName)
    {
        hMutex = CreateMutex(NULL, FALSE, szMutexName);
        WaitForSingleObject(hMutex, INFINITE);
    }

    inline CLock::~CLock()
    {
        ReleaseMutex(hMutex);
        CloseHandle(hMutex);
    }

    class CBankAccount;
    ```

```cpp
class CParameters
{
public:
    CParameters(CBankAccount* fromAccount, CBankAccount* toAccount,
        double dAmount, bool bPrintOutput = true)
    {
        this->fromAccount = fromAccount;
        this->toAccount = toAccount;
        this->dAmount = dAmount;
        this->bPrintOutput = bPrintOutput;
    }
    CBankAccount* fromAccount;
    CBankAccount* toAccount;
    double dAmount;
    bool bPrintOutput;
};

class CBankAccount
{
public:
    CBankAccount(double dBalance) : dBalance(dBalance) {
        uID = NewId(); *szLock = 0; _tcscpy_s(szLock, LockName());
    }
    CBankAccount(double dBalance, TCHAR* szLockName) :
        dBalance(dBalance) {
        uID = NewId(); *szLock = 0; _tcscpy_s(szLock, szLockName);
    }
    static DWORD WINAPI Transfer(LPVOID lpParameter);
    double& Balance() { return dBalance; }
    unsigned AccountID() const { return uID; }
    TCHAR* LockName(void);
private:
    unsigned uID;
    double dBalance;
    TCHAR szLock[32];
    static unsigned NewId() {
        static unsigned uSeed = 61524; return
            uSeed++;
    }
};

TCHAR* CBankAccount::LockName(void)
{
    static int iCount = 0;
    static TCHAR szBuffer[32];

    if (*szLock == 0)
    {
        wsprintf(szBuffer, _T("_lock_%d_"), ++iCount);
    }
    else
    {
        return szLock;
    }

    return szBuffer;
}
```

```
DWORD WINAPI CBankAccount::Transfer(LPVOID lpParameter)
{
    CParameters* parameters = (CParameters*)lpParameter;

    CLock* outerLock = new CLock(parameters->fromAccount->szLock);

    if (parameters->dAmount < parameters->fromAccount->dBalance)
    {
        CLock* innerLock = new CLock(parameters->toAccount->szLock);
        parameters->fromAccount->dBalance -= parameters->dAmount;
        parameters->toAccount->dBalance += parameters->dAmount;
        delete innerLock;
        delete outerLock;
        if (parameters->bPrintOutput)
        {
            _tprintf_s(_T("%ws\n%ws\t\t%8.2lf\n%ws\t%8u\n%ws\t%8.2lf\n%ws\t%8u\n%ws\t%8.2lf\n\n"),
                L"Transfer succeeded.",
                L"Amount:", parameters->dAmount,
                L"From account:", parameters->fromAccount->AccountID(),
                L"Balance:", parameters->fromAccount->Balance(),
                L"To account:", parameters->toAccount->AccountID(),
                L"Balance:", parameters->toAccount->Balance());
        }

        delete parameters;
        return 1;
    }
    delete outerLock;

    if (parameters->bPrintOutput)
    {
        _tprintf_s(_T("%ws\n%ws\t\t%8.2lf\n%ws\t%8u\n%ws\t%8.2lf\n%ws\t%8u\n%ws\t%8.2lf\n%ws\n\n"),
            L"Transfer failed.",
            L"Amount:", parameters->dAmount,
            L"From account:", parameters->fromAccount->AccountID(),
            L"Balance:", parameters->fromAccount->Balance(),
            L"To account:", parameters->toAccount->AccountID(),
            L"Balance:", parameters->toAccount->Balance(),
            L"Not enough funds!");
    }

    delete parameters;
    return 0;
}
#endif
```

4. 打开【解决方案资源管理器】，右键单击【头文件】，添加一个新的头文件 CThread。打开 CThread.h，输入下面的代码：

```
#ifndef __THREAD__
#define __THREAD__

#include "main.h"

class CThread
{
```

```cpp
public:
    CThread(LPTHREAD_START_ROUTINE lpThreadStart) : dwExitCode(0),
        lpThreadStart(lpThreadStart) { }
    ~CThread();
    void Start(LPVOID lpContext = 0);
    bool StillAlive() { return ExitStatus() == STILL_ACTIVE; }
    HANDLE Handle() const { return hThread; }
    DWORD ThreadId() const { return dwThreadId; }
    DWORD ExitStatus() {
        GetExitCodeThread(hThread, &dwExitCode);
        return dwExitCode;
    }
    static DWORD GetThreadId(CThread* cThread);
private:
    LPTHREAD_START_ROUTINE lpThreadStart;
    HANDLE hThread;
    DWORD dwThreadId;
    DWORD dwExitCode;
};

CThread::~CThread()
{
    if (StillAlive())
    {
        TerminateThread(hThread, (DWORD)-1);
    }
    CloseHandle(hThread);
}

void CThread::Start(LPVOID lpContext)
{
    hThread = CreateThread(NULL, 0, lpThreadStart, lpContext, 0, &dwThreadId);
}

DWORD CThread::GetThreadId(CThread* cThread)
{
    return cThread->ThreadId();
}

#endif
```

5. 把第 1 章中的 CList.h 头文件复制到该项目的目录中。打开【解决方案资源管理器】,右键单击【头文件】,选择【添加】-【现有项】,添加 CList.h。

6. 打开【解决方案资源管理器】,右键单击【头文件】,添加一个新的头文件 CThreadPool。打开 CThreadPool.h。输入下面的代码:

```cpp
#ifndef __THREADPOOL__
#define __THREADPOOL__

#include "main.h"
#include "CList.h"
#include "CThread.h"

class CThreadPool
{
public:
```

```cpp
    CThreadPool();
    CThreadPool(unsigned uMaxThreads) : dwMaxCount(uMaxThreads),
        threadList(new CList<CThread>()) { }
    ~CThreadPool() { RemoveAll(); }
    DWORD Count() { return threadList->Count(); }
    void RemoveThread(DWORD dwThreadId);
    void WaitAll();
    void RemoveAll();
    DWORD& MaxCount() { return dwMaxCount; }
    void ReleaseThread(DWORD dwThreadId);
    CThread* RequestThread(LPTHREAD_START_ROUTINE threadStart);
private:
    CList<CThread>* threadList;
    DWORD dwMaxCount;
};

CThreadPool::CThreadPool()
{
    SYSTEM_INFO sysInfo;
    GetSystemInfo(&sysInfo);
    dwMaxCount = sysInfo.dwNumberOfProcessors;

    threadList = new CList<CThread>();
}

void CThreadPool::RemoveThread(DWORD dwThreadId)
{
    CThread* thread = threadList->Find(CThread::GetThreadId,
        dwThreadId);
    if (thread)
    {
        delete thread;
    }
}

void CThreadPool::WaitAll()
{
    HANDLE* hThreads = new HANDLE[threadList->Count()];
    CThread* thread = 0;

    for (unsigned uIndex = 0; uIndex < threadList->Count();
    uIndex++)
    {
        thread = threadList->GetNext(thread);
        hThreads[uIndex] = thread->Handle();
    }

    WaitForMultipleObjects(threadList->Count(), hThreads, TRUE,
        INFINITE);
    delete[] hThreads;
}

void CThreadPool::RemoveAll()
{
    CThread* thread = 0;
    while (thread = threadList->GetNext(thread))
    {
        threadList->Remove(thread);
        thread = 0;
```

```cpp
        }
        delete threadList;
    }

    void CThreadPool::ReleaseThread(DWORD dwThreadId)
    {
        CThread* thread = threadList->Find(CThread::GetThreadId,
            dwThreadId);
        if (thread != NULL)
        {
            TerminateThread(thread->Handle(), (DWORD)-1);
        }
    }

    CThread* CThreadPool::RequestThread(LPTHREAD_START_ROUTINE
        lpThreadStart)
    {
        CThread* thread = NULL;
        if (Count() < MaxCount())
        {
            threadList->Insert(thread = new CThread(lpThreadStart));
            return thread;
        }

        while (thread = threadList->GetNext(thread))
        {
            if (!thread->StillAlive())
            {
                break;
            }
            Sleep(100);
        }

        if (thread == NULL)
        {
            thread = threadList->GetFirst();
        }

        threadList->Remove(thread);

        threadList->Insert(thread = new CThread(lpThreadStart));

        return thread;
    }

    #endif
```

7. 打开【解决方案资源管理器】,右键单击【源文件】,添加一个新的源文件 main。打开 main.cpp,输入下面的代码:

```cpp
    #include "main.h"
    #include "CBankAccount.h"
    #include "CThreadPool.h"

    double Random(double dMin, double dMax)
    {
        double dValue = (double)rand() / RAND_MAX;
        return dMin + dValue * (dMax - dMin);
```

```
}

int main(void)
{
    srand((unsigned)time(NULL));

#define ACCOUNTS_COUNT 12
    CBankAccount* accounts[ACCOUNTS_COUNT];
    for (unsigned uIndex = 0; uIndex < ACCOUNTS_COUNT; uIndex++)
    {
        accounts[uIndex] = new CBankAccount(Random(10, 1000));
    }

    CThreadPool* pool = new CThreadPool();

#define TASK_COUNT 5
    for (unsigned uIndex = 0; uIndex < TASK_COUNT; uIndex++)
    {
        int nFirstIndex = rand() % ACCOUNTS_COUNT;
        int nSecondIndex = -1;
        while ((nSecondIndex = rand() % ACCOUNTS_COUNT) ==
            nFirstIndex) {}

        CParameters* params = new CParameters(accounts[nFirstIndex],
            accounts[nSecondIndex], Random(50, 200));
        pool->RequestThread(CBankAccount::Transfer)->Start(params);
    }

    pool->WaitAll();
    delete pool;

#ifdef _DEBUG
    return system("pause");
#endif
    return 0;
}
```

## 示例分析

首先，解释一下头文件。在 main.h 头文件中，包含了必需的 Windows 头文件。在 CBankAccount.h 中创建了 CBankAccount 类，作为银行账户的抽象。实现了两个辅助类：CLock 和 CParameters。CLock 类实现一个简单的锁机制，提供 CLock 对象从被创建到被销毁期间的独占访问。CParameters 类供传递给工作线程的参数使用。

CBankAccount 类实现银行账户对象。该类包含多个特征：uID 用于表示银行账户的 ID；dBalance 用于表示资金数量；szLock 用于表示锁名。该类有两个构造函数：第 1 个构造函数只能设置账户余额，使用 LockName 方法生成锁名；第 2 个构造函数可以设置账户余额和锁名。另外，还有两个静态方法：NewId 和 Transfer。NewId 方法用于生成账户 ID，Transfer 方法用于从一个账户把资金转移到另一个账户。LockName 方法可用于从任意银行账户获取当前的锁名。Transfer 方法先从第 1 个账户（fromAccount）中获得一个锁，再从在第 2 个账户（toAccount）中获得另一个锁。当两个账户都被锁定后，才能执行转账，把从 fromAccount 减去的资金数额加到 toAccount 中。如果转账成功，该方法将返回 1；如果转账失败

（如果帐户余额少于要转出的数额），则返回 0。

在 CThread.h 头文件中，实现了 CTread 类。CThread 对象将用作线程池中工作项的抽象。该类包含 lpThreadStart 特征，用来表示线程的开始地址；hThread 用来表示线程句柄；dwThreadId 用来表示线程 ID；dwExitCode 用来表示线程退出码。该类实现了一个构造函数，当 lpThreadStart 指向用户提供的开始地址时，退出码被设置为零。在析构函数中，我们通过 StillAlive 方法询问线程是否仍然活跃，如果活跃则调用 TerminateThread API，然后关闭该线程的句柄。Start 方法创建线程，并给线程的开始地址传递一个参数（如果有的话）。StillAlive 方法比较 STILL_ACTIVE 宏和 ExitStatus 的返回值，并返回**布尔值**。Handle 和 ThreadId 方法分别返回线程的句柄和线程 ID。ExitStatus 方法调用 GetExitCodeThread API 获取退出码或 STILL_ACTIVE。

在 CThreadPool.h 头文件中，CThreadPool 类实现真正的线程池。该类有两个特征：一个是指向工作项列表的指针；另一个是线程池可创建的最大线程数量。该类有两个构造函数：第 1 个构造函数没有参数，默认用机器的逻辑处理器数量设置线程的最大计数；第 2 个构造函数有一个参数，即线程池可实例化的实际最大线程数量。在析构函数中，调用 RemoveAll 方法释放用过的资源。Count 方法返回线程列表上的项数。RemoveThread 方法查找 dwThreadId 参数指定的线程。如果找到，就移除它。WaitAll 方法创建一个线程句柄数组，WaitForMultipleObjects API 中要用到该数组。为了等待所有的线程结束它们的任务，我们把该 API 的 BOOL bWaitAll 形参指定为 TRUE。待 WaitForMultipleObjects 方法返回后，将删除线程句柄数组。RemoveAll 方法为列表中的每个线程调用 RemoveThread 方法，移除列表中的所有线程。MaxCount 方法返回 dwMaxCount 特征的引用，用于获取和设置值。如有需要，可以增减线程池的大小。ReleaseThread 方法查找 dwThreadId 参数指定的线程。如果找到，就终止线程。RequestThread 方法用于获取空闲线程。如果未达到线程的最大数量，该方法创建一个新的工作项，并将其插入线程列表中。如果已经达到最大线程数量，该方法遍历整个列表，并询问每个线程的退出码。

如果所有线程都是激活的，它将等待 0.1 秒，并再次尝试，直到它发现第 1 个完成任务的线程。当找到线程时，该方法移除该线程，并用用户提供的开始地址创建一个新线程。

main.cpp 中创建了 Random 函数，获得一个 dMin~dMax 范围内的随机 double 类型值。在应用程序的入口，我们将把随机生成器设置成当前的时间，并创建 12 个银行账户，分别用随机生成的值作为各账户的余额。然后，用待处理的任务实例化线程池。对于每个任务，我们将先后随机选择两个不同的银行账户。找到两个账户后，使用 RequestThread 方法和 Start 方法执行转账，数额为随机值。设置好所有任务后，调用 WaitAll 方法，等待任务完成。当 WaitAll 返回后，才可以删除线程池。

## 更多讨论

即使严谨地设计了线程池，注意了所有必要的细节，诸如死锁这样的情况也在所难免。对于转账的例子，账户 A 转账到账户 B 和账户 B 转账到账户 A 必须同时发生才会发生这种情况。虽然几率非常小，但还是可能发生。下一个示例将扩展该线程池,让线程池更智能,能解决诸如死锁这种问题。

## 8.3 定制线程池分发器

上一节我们为了管理某些任务(从一个银行账户转移资金到另一个银行账户)创建了一个线程池。我们的目标是使用足够的系统资源,但是也不要用得太多。同时,必须注意**繁忙**和**可用**的线程以及维护同步。下面的示例将扩展上一节实现的线程池,让它更加智能,有能力解决执行期间可能发生的问题。

这次,我们用消息传递来确定第 1 个完成任务的线程,并通知分发器,而不是用遍历线程列表并等待线程完成执行。为了更好地控制任务执行,我们将再添加一个抽象层,而不是简单地执行一个作为线程开始地址被传入的用户定义函数。

新方案的思路如图 8.2 所示。

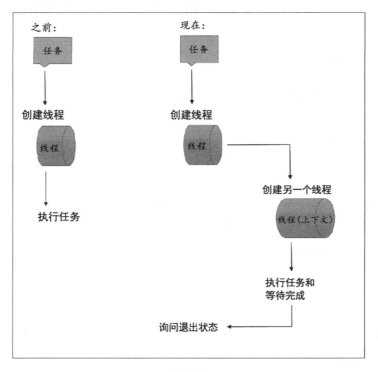

图 8.2

即使为每个任务复制若干线程,我们将创建一个线程的两个实例,仍然要维护那么多数量的线程。为执行任务创建的线程将再创建一个线程,并等待它,仅为了更好地控制任务执行。我们将调用第一个线程 `worker`,然后调用第二个线程,该线程将创建 `context thread`。这种方案背后的思想是,我们有了更多的控制。如果终止上下文线程,等待该线程的 worker 将询问它的退出码,并通知分发器:有地方出错了。或者,如果上下文线程挂起或未响应,worker 将注意到有地方出错了,并在再次通知分发器时执行相应的动作。这种粒度(*granularity*)给了我们更多的控制,而且让线程池更智能。

下面的示例中故意设计发生死锁的情形,我们将尝试解决这个问题。

### 准备就绪

确定安装并运行了 Visual Studio。

### 操作步骤

现在,根据下面的步骤创建程序,稍后再详细解释。

1. 创建一个新的空 C++控制台应用程序,并命名为 CThreadPool。

2. 打开【解决方案资源管理器】,右键单击【头文件】。添加一个新的头文件 main。打开 main.h,输入下面的代码:

    ```
    #ifndef __MAIN__
    #define __MAIN__

    #include <Windows.h>
    #include <tchar.h>
    #include <time.h>

    #define MSG_ENDTASK 0x1000
    #define MAX_WAIT_TIME 120 * 1000

    #endif
    ```

3. 打开【解决方案资源管理器】。右键单击【头文件】。添加一个新的头文件 CBankAccount。打开 CBankAccount.h,输入下面的代码:

    ```
    #ifndef __BANKACCOUNT__
    #define __BANKACCOUNT__

    #include "main.h"

    class CLock
    {
    public:
        CLock(TCHAR* szMutexName);
        ~CLock();
    private:
        HANDLE hMutex;
    };

    inline CLock::CLock(TCHAR* szMutexName)
    {
        hMutex = CreateMutex(NULL, FALSE, szMutexName);
        WaitForSingleObject(hMutex, INFINITE);
    ```

```cpp
}

inline CLock::~CLock()
{
    ReleaseMutex(hMutex);
    CloseHandle(hMutex);
}

class CBankAccount;

class CParameters
{
public:
    CParameters(CBankAccount* fromAccount, CBankAccount* toAccount,
        double dAmount, bool bPrintOutput = true)
    {
        this->fromAccount = fromAccount;
        this->toAccount = toAccount;
        this->dAmount = dAmount;
        this->bPrintOutput = bPrintOutput;
    }
    CBankAccount* fromAccount;
    CBankAccount* toAccount;
    double dAmount;
    bool bPrintOutput;
};

class CBankAccount
{
public:
    CBankAccount(double dBalance) : dBalance(dBalance) {
        uID = NewId(); *szLock = 0; _tcscpy_s(szLock, LockName());
    }
    CBankAccount(double dBalance, TCHAR* szLockName) :
        dBalance(dBalance) {
        uID = NewId(); *szLock = 0; _tcscpy_s(szLock, szLockName);
    }
    static DWORD WINAPI Transfer(LPVOID lpParameter);
    double& Balance() { return dBalance; }
    unsigned AccountID() const { return uID; }
    TCHAR* LockName(void);
private:
    unsigned uID;
    double dBalance;
    TCHAR szLock[32];
    static unsigned NewId() {
        static unsigned uSeed = 61524;
        return uSeed++;
    }
};

TCHAR* CBankAccount::LockName(void)
{
    static int iCount = 0;
    static TCHAR szBuffer[32];
    if (*szLock == 0)
    {
        wsprintf(szBuffer, _T("_lock_%d_"), ++iCount);
    }
```

```cpp
        else
        {
            return szLock;
        }
        return szBuffer;
    }

    DWORD WINAPI CBankAccount::Transfer(LPVOID lpParameter)
    {
        CParameters* parameters = (CParameters*)lpParameter;

        Sleep(100);

        CLock* outerLock = new CLock(parameters->fromAccount->szLock);

        if (parameters->dAmount < parameters->fromAccount->dBalance)
        {
            CLock* innerLock = new CLock(parameters->toAccount->szLock);
            parameters->fromAccount->dBalance -= parameters->dAmount;
            parameters->toAccount->dBalance += parameters->dAmount;
            delete innerLock;
            delete outerLock;

            if (parameters->bPrintOutput)
            {
                _tprintf_s(_T("%ws\n%ws\t\t%8.2lf\n%ws\t%8u\n%ws
                    \t%8.2lf\n%ws\t % 8u\n%ws\t%8.2lf\n\n"),
                    L"Transfer succeeded.",
                    L"Amount:", parameters->dAmount,
                    L"From account:", parameters->fromAccount->AccountID(),
                    L"Balance:", parameters->fromAccount->Balance(),
                    L"To account:", parameters->toAccount->AccountID(),
                    L"Balance:", parameters->toAccount->Balance());
            }

            delete parameters;
            return 1;
        }
        delete outerLock;

        if (parameters->bPrintOutput)
        {
            _tprintf_s(_T("%ws\n%ws\t\t%8.2lf\n%ws\t%8u\n%ws
                \t%8.2lf\n%ws\t % 8u\n%ws\t%8.2lf\n%ws\n\n"),
                L"Transfer failed.",
                L"Amount:", parameters->dAmount,
                L"From account:", parameters->fromAccount->AccountID(),
                L"Balance:", parameters->fromAccount->Balance(),
                L"To account:", parameters->toAccount->AccountID(),
                L"Balance:", parameters->toAccount->Balance(),
                L"Not enough funds!");
        }

        delete parameters;
        return 0;
    }

#endif
```

4. 打开【解决方案资源管理器】，右键单击【头文件】，添加一个新的头文件 CThread。打开 CThread.h，输入下面的代码：

```
#ifndef __THREAD__
#define __THREAD__

#include "main.h"

class CThread
{
public:
    CThread(LPTHREAD_START_ROUTINE lpThreadStart, DWORD
        dwDispatcherId);
    ~CThread();
    void Start(LPVOID lpContext = 0) {
        pThreadContext->lpParameter = lpContext; ResumeThread(hThread);
    }
    bool StillAlive() { return ExitStatus() == STILL_ACTIVE; }
    HANDLE Handle() const { return hThread; }
    HANDLE ContextHandle() const { return pThreadContext->hThread; }
    DWORD ThreadId() const { return dwThreadId; }
    DWORD ContextThreadId() const { return pThreadContext->dwThreadId; }
    DWORD ExitStatus() {
        GetExitCodeThread(hThread, &dwExitCode);
        return dwExitCode;
    }
    CThread* SetMaxWaitTime(DWORD dwMilliseconds) {
        pThreadContext->dwMaxWaitTime = dwMilliseconds;
        return this;
    }
    static DWORD GetThreadId(CThread* cThread);
protected:
    static DWORD WINAPI StartAddress(LPVOID lpParam);
private:
    typedef struct
    {
        LPTHREAD_START_ROUTINE lpThreadStart;
        LPVOID lpParameter;
        HANDLE hThread;
        DWORD dwThreadId;
        DWORD dwDispatcherId;
        DWORD dwMaxWaitTime;
    } STARTCONTEXT, *PSTARTCONTEXT;
    HANDLE hThread;
    DWORD dwThreadId;
    DWORD dwExitCode;
    PSTARTCONTEXT pThreadContext;
};

CThread::CThread(LPTHREAD_START_ROUTINE lpThreadStart, DWORD
    dwDispatcherId) : dwExitCode(0)
{
    pThreadContext = new STARTCONTEXT();
    pThreadContext->lpThreadStart = lpThreadStart;
    pThreadContext->dwDispatcherId = dwDispatcherId;
    pThreadContext->dwMaxWaitTime = MAX_WAIT_TIME;
    hThread = CreateThread(NULL, 0, StartAddress, this,
        CREATE_SUSPENDED, &dwThreadId);
```

```
}
CThread::~CThread()
{
    delete pThreadContext;
    if (StillAlive())
    {
        TerminateThread(hThread, (DWORD)-1);
    }
    CloseHandle(hThread);
}
DWORD CThread::GetThreadId(CThread* cThread)
{
    return cThread->ThreadId();
}
DWORD WINAPI CThread::StartAddress(LPVOID lpParam)
{
    CThread* thread = (CThread*)lpParam;
    thread->pThreadContext->hThread = CreateThread(NULL, 0,
        thread->pThreadContext->lpThreadStart,
        thread->pThreadContext->lpParameter, 0,
        &thread->pThreadContext->dwThreadId);

    DWORD dwStatus = WaitForSingleObject(
        thread->pThreadContext->hThread,
        thread->pThreadContext->dwMaxWaitTime);

    CloseHandle(thread->pThreadContext->hThread);

    PostThreadMessage(thread->pThreadContext->dwDispatcherId,
        MSG_ENDTASK, (WPARAM)thread->dwThreadId, 0);

    return 0L;
}

#endif
```

5. 把第 1 章中的 CList.h 头文件复制到该项目的目录中。打开【解决方案资源管理器】，右键单击【头文件】，选择【添加】-【现有项】，添加 CList.h。

6. 打开【解决方案资源管理器】，右键单击【头文件】，添加一个新的头文件 CThreadPool。打开 CThreadPool.h。输入下面的代码：

```
#ifndef __THREADPOOL__
#define __THREADPOOL__

#include "main.h"
#include "CList.h"
#include "CThread.h"
#include <Wct.h>

#pragma comment (lib, "Advapi32.lib")

class CThreadPool
{
public:
    CThreadPool();
```

```cpp
        CThreadPool(unsigned uMaxThreads);
        ~CThreadPool();
        DWORD Count() { return threadList->Count(); }
        void RemoveThread(DWORD dwThreadId);
        void WaitAll();
        void RemoveAll();
        DWORD& MaxCount() { return dwMaxCount; }
        void ReleaseThread(DWORD dwThreadId);
        CThread* RequestThread(LPTHREAD_START_ROUTINE threadStart);
    private:
        void InitializePool();
        static DWORD WINAPI ProblemSolver(LPVOID lpParam);
        void ClearMessageQueue() {
            MSG msg;
            while (PeekMessage(&msg, NULL, 0, 0, PM_REMOVE));
        }
        HANDLE hProblemSolver;
        HANDLE hEvent;
        CList<CThread>* threadList;
        DWORD dwMaxCount;
};

CThreadPool::CThreadPool()
{
    SYSTEM_INFO sysInfo;
    GetSystemInfo(&sysInfo);
    dwMaxCount = sysInfo.dwNumberOfProcessors;

    InitializePool();
}

CThreadPool::CThreadPool(unsigned uMaxThreads)
{
    dwMaxCount = uMaxThreads;

    InitializePool();
}

CThreadPool::~CThreadPool()
{
    SetEvent(hEvent);

    if (WaitForSingleObject(hProblemSolver, MAX_WAIT_TIME) != WAIT_OBJECT_0)
    {
        TerminateThread(hProblemSolver, (DWORD)-1);
    }

    CloseHandle(hProblemSolver);

    CloseHandle(hEvent);

    RemoveAll();
    ClearMessageQueue();
}

void CThreadPool::RemoveThread(DWORD dwThreadId)
{
    CThread* thread = threadList->Find(CThread::GetThreadId, dwThreadId);
    if (thread)
```

```cpp
        {
            delete thread;
        }
}

void CThreadPool::WaitAll()
{
    HANDLE* hThreads = new HANDLE[threadList->Count()];
    CThread* thread = 0;

    for (unsigned uIndex = 0; uIndex < threadList->Count();
    uIndex++)
    {
        thread = threadList->GetNext(thread);
        hThreads[uIndex] = thread->Handle();
    }

    WaitForMultipleObjects(threadList->Count(), hThreads, TRUE, INFINITE);
    delete[] hThreads;
}

void CThreadPool::RemoveAll()
{
    CThread* thread = 0;
    while (thread = threadList->GetNext(thread))
    {
        threadList->Remove(thread);
        thread = 0;
    }
    delete threadList;
}

void CThreadPool::ReleaseThread(DWORD dwThreadId)
{
    CThread* thread = threadList->Find(CThread::GetThreadId, dwThreadId);
    if (thread != NULL)
    {
        TerminateThread(thread->Handle(), (DWORD)-1);
    }
}

CThread* CThreadPool::RequestThread(LPTHREAD_START_ROUTINE threadStart)
{
    CThread* thread = NULL;
    if (Count() < MaxCount())
    {
        threadList->Insert(thread = new CThread(threadStart,
            GetCurrentThreadId()));
        return thread;
    }

    while (thread = threadList->GetNext(thread))
    {
        if (!thread->StillAlive())
        {
            break;
        }
    }
```

## 8.3 定制线程池分发器

```cpp
            if (thread == NULL)
            {
                MSG msg = { 0 };
                while (GetMessage(&msg, NULL, 0, 0) > 0)
                {
                    thread = threadList->Find(CThread::GetThreadId,
                        (DWORD)msg.wParam);
                    if (thread)
                    {
                        break;
                    }
                }
            }

            threadList->Remove(thread);

            threadList->Insert(thread = new CThread(threadStart,
                GetCurrentThreadId()));

            return thread;
}

void CThreadPool::InitializePool()
{
    MSG msg;
    PeekMessage(&msg, NULL, WM_USER, WM_USER, PM_NOREMOVE);

    hEvent = CreateEvent(NULL, TRUE, FALSE, _T("__prbslv_1342__"));
    hProblemSolver = CreateThread(NULL, 0, ProblemSolver, this, 0, NULL);

    threadList = new CList<CThread>();
}

DWORD WINAPI CThreadPool::ProblemSolver(LPVOID lpParam)
{
    CThreadPool* pool = (CThreadPool*)lpParam;

    DWORD dwWaitStatus = 0;
    while (true)
    {
        dwWaitStatus = WaitForSingleObject(pool->hEvent, 10);
        if (dwWaitStatus == WAIT_OBJECT_0)
        {
            break;
        }

        CThread* thread = NULL;
        while (thread = pool->threadList->GetNext(thread))
        {
            if (thread->StillAlive())
            {
                HWCT hWct = NULL;
                DWORD dwNodeCount = WCT_MAX_NODE_COUNT;
                BOOL bDeadlock = FALSE;
                WAITCHAIN_NODE_INFO NodeInfoArray[WCT_MAX_NODE_COUNT];

                hWct = OpenThreadWaitChainSession(0, NULL);
                if (GetThreadWaitChain(hWct, NULL,
                    WCTP_GETINFO_ALL_FLAGS, thread->ContextThreadId(),
```

```
                        &dwNodeCount, NodeInfoArray, &bDeadlock))
                    {
                        if (bDeadlock)
                        {
                            if (TerminateThread(thread->ContextHandle(),
                                (DWORD)-1))
                            {
                                _tprintf_s(
#ifdef _UNICODE
                                    _T("%ws\n%ws\t[%u]%ws\n\n"),
#else
                                    _T("%s\n%s\t[%u]%s\n\n"),
#endif
                                    _T("Error! Deadlock found!"), _T("Thread:"),
                                        thread->ContextThreadId(),
                                    _T("terminated!"));
                            }
                        }
                    }
                    CloseThreadWaitChainSession(hWct);
                }
            }
            Sleep(1000);
        }
        return 0L;
    }

#endif
```

7. 打开【解决方案资源管理器】，右键单击【源文件】，添加一个新的源文件 main。打开 main.cpp，输入下面的代码：

```
#include "main.h"
#include "CBankAccount.h"
#include "CThreadPool.h"

int main(void)
{
    int i = 0;
    while (i++ < 10)
    {
        CBankAccount* a = new CBankAccount(200);
        CBankAccount* b = new CBankAccount(200);

        CThreadPool* pool = new CThreadPool(4);

        CParameters* params1 = new CParameters(a, b, 100);
        pool->RequestThread(
            CBankAccount::Transfer)->SetMaxWaitTime(
                6000)->Start(params1);

        CParameters* params2 = new CParameters(b, a, 50);
        pool->RequestThread(CBankAccount::Transfer)->Start(params2);

        pool->WaitAll();
        delete pool;
    }
```

```
#ifdef _DEBUG
    return system("pause");
#endif
    return 0;
}
```

## 示例分析

首先，解释一下头文件。在 main.h 中，除了包含了必需的 Windows 头文件外，还定义了两个宏：MSG_ENDTASK 用于从工人线程把完成任务的消息寄送给分发器；为了让线程不用等待它的上下文线程太长时间，MAX_WAIT_TIME 设置等待的最长时间。可以在 CThread 对象中使用 SetMaxWaitTime 方法增加或减少最长等待时间。我们稍后解释。

CBankAccount.h 中实现的 CBankAccount 类、CLock 类和 CParameters 基本不变。在 CBankAccount 类中，我们稍微改动了 Transfer 方法的开头部分，添加了 0.1 秒的睡眠时间，这样做是为了故意创造死锁的条件。启动两个线程后，如果移除睡眠时间，死锁可能不会发生，系统会给线程分配足够的 CPU 时间。

在 CThread.h 中，实现了更智能的 CThread 类。我们把线程池中的 CThread 对象作为工作对象的抽象。它有 lpThreadStart 特征，用于表示线程的开始地址；hThread 用于表示线程句柄；dwThreadId 用于表示线程 ID；dwExitCode 用于表示线程的退出码；pThreadContext 用于表示上下文线程。如图 8.2 所示，为了能更好地控制执行，我们创建了另一个线程，即上下文线程。上下文实现是由这些特征组成的一个内部结构：lpThreadStart 用于表示用户定义的实际开始地址；lpParameter 用于表示用户传递的开始地址；hThread 用于表示上下文线程句柄；dwThreadId 用于表示上下文线程 ID；dwDispatcherId 用于表示线程池管理者（分发器）的 ID；dwMaxWaitTime 用于设置等待上下文线程的超时时间。CThread 类实现了一个构造函数，其退出码设置为零。构造函数的参数也改动了，为了给线程传递实际的管理者标识符，我们在原来 lpThreadStart 的基础上，又添加了 dwDispatcherId。在构造函数中，我们创建了一个 STARTCONTEXT 对象，分别用传入的 lpThreadStart 和 dwDispatcherId 设置用户定义的开始地址和线程池管理者 ID；用 MAX_WAIT_TIME 设置 dwMaxWaitTime；还创建了一个工人线程，以 CREATE_SUSPENDED 标志作为其中的一个参数。在析构函数中，我们首先删除 pThreadContext 用过的内存。然后，询问线程是否活跃，如果活跃，则调用 TerminateThread API，然后关闭线程句柄。Start 方法设置用户传递的 lpParameter 特征，并恢复工人线程。StillAlive 方法比较 STILL_ACTIVE 宏和 ExitStatus 的返回值，并返回布尔值。Handle 和 ThreadId 方法分别返回工人线程的句柄和工人线程 ID。ContextHandle 和 ContextThreadId 方法分别返回上下文线程句柄和上下文 ID。ExitStatus 方法调用 GetExitCodeThread API 以获得线程的退出码或 STILL_ACTIVE。SetMaxWaitTime 方法设置最长等待时间特征值。我们添加了 StartAddress 方法，用于通过创建另一个线程（即上下文线程）来执行任务，并等待它完成任务（在最糟糕的情况中，等待时间间隔为 pThreadContext->dwMaxWaitTime 的最大值）。要么等待经过时间间隔，要么返回上下文线程。然后，上下文线程句柄被关闭，MSG_ENDTASK

消息被寄送给分发器消息队列。

`CParameters` 类仍然相同。我们没有在指定等待时间内遍历线程列表，而是阻塞分发器并等待第 1 个可用的工人线程寄送的 `MSG_ENDTASK` 消息。

在 `CThreadPool` 类中，包含了新增的头文件 `Wct.h`。为了使用与线程等待链相关的 API，还添加了 `Advapi32` 库。我们稍后解释。`CThreadPool` 类有 4 个特征：`hProblemSolver` 线程的句柄，用于偶尔检查潜在的问题；手动重置事件的 `hEvent` 句柄，用于在问题解决器线程退出时报警；指向工作项列表的 `threadList` 指针；`dwMaxCount` 用于表示线程池可创建的最大线程数量。该类实现了两个构造函数，第 1 个构造函数默认使用机器逻辑处理器数量设置最大数量的线程；第 2 个构造函数只有一个参数，即线程池可实例化的实际最大线程数量。两个构造函数都调用 `InitializePool` 方法为分发器创建消息队列，另外还创建一个事件、一个问题解决器线程和一个新的线程列表。在析构函数中，我们首先把事件设置为 `signaled` 状态，并等待 `MAX_WAIT_TIME` 时长安全退出问题解决器线程。如果最长等待时间过去了还没有线程完成任务，就使用 `TerminateThread` API 强制结束线程的执行。然后关闭两个句柄，我们使用 `RemoveAll` 方法释放用过的资源，用 `ClearMessageQueue` 方法清空消息队列。

`Count` 方法返回线程列表中工作项的计数。`RemoveThread` 方法查找 `dwThreadId` 参数指定的线程，如果找到，就返回该线程。`WaitAll` 方法创建一个线程句柄数组，`WaitForMultipleObjects` API 中要用到该数组。为了等待所有的线程结束它们的任务，我们把该 API 的 `BOOL bWaitAll` 形参指定为 `TRUE`。待 `WaitForMultipleObjects` 方法返回后，将删除线程句柄数组。`RemoveAll` 方法为线程列表中的每个线程调用 `RemoveThread` 方法，移除所有的线程。`MaxCount` 方法返回 `dwMaxCount` 特征的引用，用于获得或设置它的值。再次强调，如果需要可以增减线程池的大小。`ReleaseThread` 方法查找 `dwThreadId` 参数指定的线程，如果找到，终止该线程。

`RequestThread` 方法用于获得空闲线程。如果未达到线程的最大数量，该方法将创建一个新的工作项，并将其插入线程队列。如果达到了线程最大数量，`RequestThread` 将遍历整个线程列表，询问每个线程的退出码。如果所有的线程都是激活的，分发器将使用 `GetMessage` API 进行阻塞，并在第 1 个可用的工人线程寄送完成任务消息之前一直等待。然后，我们将用传入的线程 ID 查找线程，如果找到了就移除它，并用用户提供的开始地址创建一个新线程。`ProblemSolver` 方法用作每个线程池的监视例程。

如图 8.3 所示，把问题解决器作为另一个并行执行的线程非常重要。

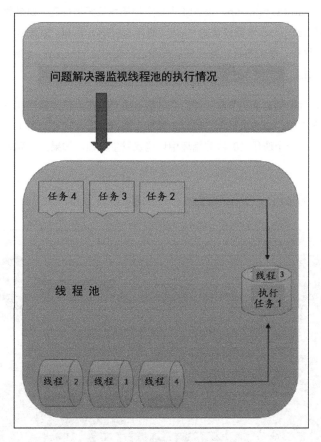

图 8.3

如果因某种原因而出现问题导致线程池被阻塞，问题解决器线程将尝试解决这个问题。因此，不能把该线程放在线程池中的任何线程中执行，否则它会和线程池一起被阻塞。它的任务是无限循环，或者在程序退出（事件被设置为 signaled 状态）前一直循环。所以，每次迭代都将调用 WaitForSingleObject，它的两个参数是传入的事件句柄和 0.01 秒等待间隔。如果事件被设置为 signaled 状态，线程将退出。否则，它将遍历整个线程列表，为每个线程调用 StillAlive 方法。如果线程是激活的，它将调用 OpenThreadWaitChainSession API 创建等待链遍历（WCT）。

 等待链遍历（*Wait Chain Traversal*，WCT）能诊断最终的应用程序挂起和死锁。等待链是线程和同步对象的交替序列（*alternating sequence*）；等待链中的每个线程都等待紧随其后的同步对象，而该同步对象又归其后的线程所有（摘自 MSDN）

接着，调用 `GetThreadWaitChain`，该 API 为指定的线程检索等待链。如果数组中节点的任何子集形成一个循环，该函数将把 `IsCycle` 参数设置为 TRUE（摘自 MSDN）。所以，如果形成一个循环，dDeadlock 将被设置为 TRUE。如果这样，我们将终止被死锁的线程，并使用 `CloseThreadWaitChainSession` API 关闭 WCT 部分。然后，挂起该线程 1 秒钟，并继续监视线程池。

在应用程序的入口，我们创建了两个有一些资金的银行账户。在创建好线程池后，请求线程把第 1 个账户的资金转移到第 2 个账户中，并请求另一个线程把第 2 个账户的资金转移到第 1 个账户中，试图制造有问题的情况。而且，将其放在一个迭代 10 次的循环中以确保发生死锁。如果出现死锁，问题解决器就会起作用。

带死锁的情况如图 8.4 所示。

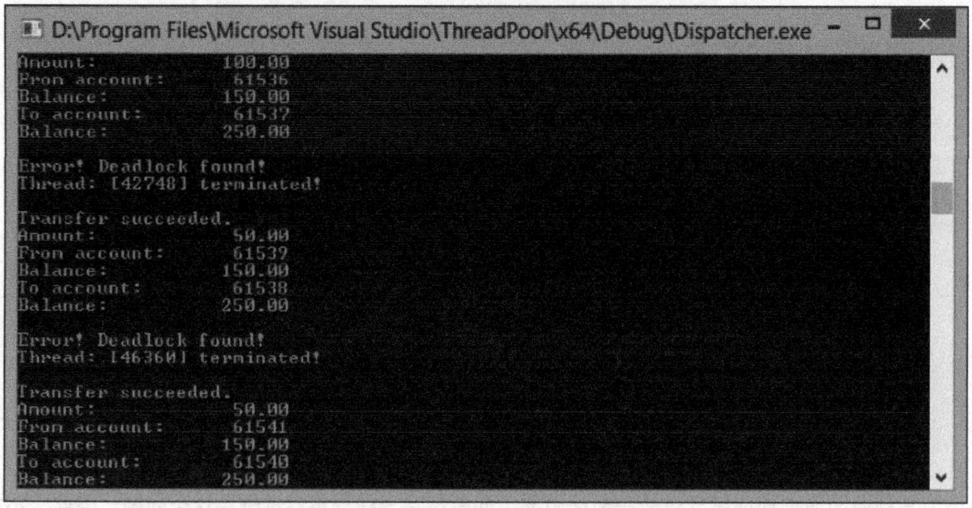

图 8.4

## 更多讨论

以上示例是为了演示某种可能出状况的情形。这种情况被称为**银行家算法**（*Banker's algorithm*），由 Edsger Dijkstra 设计。该算法与上面的线程池实现和解决方案非常像，可用于任何有死锁或类似问题发生的情形。我们的初衷是告诉读者，谨慎地设计对象不仅让程序的开发过程更加容易，而且能带来诸多便利，如控制程序的执行、妥善处理已使用的资源、使用和执行高级管理，以及监视程序执行。一般而言（我们提到很多次了），谨慎设计应用程序非常重要，如有必要，在设计中加入高级机制支持。

## 8.4 使用远程线程

到目前为止，我们着重介绍的是，完全控制应用程序的各部分和程序的执行。在现实中，远远不是这样。你可能要经常接着别人的工作继续开发，或者只能使用已编译的部分，甚至是最终应用程序（在里面改变代码是不可能的）。尽管如此，有时也需要改变应用程序的某些行为。我们稍后解释。

当我们不能改代码时，可以使用**远程线程**（*remote threading*）。远程线程是 Windows 操作系统从 XP 开始提供的一种特性，在后续版本中一直存在。实际上，远程线程与其他线程一样，也是线程。区别在于远程线程要在远程进程的空间地址中执行，不能在当前进程上下文中执行。这一特性非常适合处理我们之前描述的情况。

启动远程线程的方式有多种：使用 `CreateRemoteThread` API、`SetWindowsHookEx` API 或者 `NtCreateThreadEx` API。

我们重点介绍 `CreateRemoteThread` API。在下面的示例中，假设 Windows 计算器（`calc.exe`）是系统的内部应用程序，我们不能更改里面的代码。但是，我们必须更改它的主窗口标题（出于某种原因）。其实，要更改计算器窗口的标题没必要使用远程线程，但是为了演示远程线程的用法，我们还是小题大做一回。

### 准备就绪

确定安装并运行了 Visual Studio。

### 操作步骤

现在，根据下面的步骤创建程序，稍后再详细解释。

1. 创建一个新的空 C++ 控制台应用程序，并命名为 `RemoteThreading`。

2. 打开【窗口资源管理器】，导航至解决方案文件夹。创建一个新的文本文件 `common.txt`。改变它的扩展名为 `.h`。

3. 打开【解决方案资源管理器】，右键单击【远程线程】解决方案。选择【添加】-【现有项】，添加 `common.h`。打开 `common.h`，输入下面的代码：

```
#include <windows.h>
#include <tchar.h>

#define MAPPING_NAME _T("__comm_61524_map__")
#define EVENT_NAME   _T("__evnt_68435_rst__")

typedef struct
{
    HMODULE hLibrary;
```

```
} COMM_OBJECT, *PCOMM_OBJECT;
```

4. 打开【解决方案资源管理器】,在【RemoteThreading】项目下面右键单击【源文件】。添加一个新的源文件 main。打开 main.cpp。

5. 输入下面的代码:

```
#include <windows.h>
#include <tchar.h>
#include "..\common.h"

int __stdcall RemoteLoadLibrary(HANDLE hProcess, char*
    szLibraryName)
{
    LPTHREAD_START_ROUTINE lpLoadLibrary = (LPTHREAD_START_ROUTINE)
        GetProcAddress(GetModuleHandleA("Kernel32.dll"), "LoadLibraryA");
    if (lpLoadLibrary == NULL)
    {
        return -1;
    }

    if (hProcess == NULL)
    {
        _tprintf_s(_T("Handle to Process 0x%x was NULL!\n"), (int)hProcess);
        return -1;
    }

    size_t uMemSize = strlen(szLibraryName) + 1;
    void* lpRemoteMem = VirtualAllocEx(hProcess, NULL, uMemSize,
        MEM_COMMIT, PAGE_READWRITE);
    if (lpRemoteMem == NULL)
    {
        _tprintf_s(
#ifdef _UNICODE
            _T("%ws\nError:\t%u\n"),
#else
            _T("%s\nError:\t%u\n"),
#endif
            _T("Could not allocate remote virtual memory!"),
                GetLastError());
        return -1;
    }

    BOOL bSuccess = WriteProcessMemory(hProcess, lpRemoteMem,
        szLibraryName, uMemSize, NULL);
    if (bSuccess == STATUS_ACCESS_VIOLATION || bSuccess == FALSE)
    {
        _tprintf_s(
            _T("Could not write remote virtual memory!\nError:\t%u\n"),
                GetLastError());
        VirtualFreeEx(hProcess, lpRemoteMem, 0, MEM_RELEASE);
        return -1;
    }

    HANDLE hThread = CreateRemoteThread(hProcess, NULL, 0,
        lpLoadLibrary, lpRemoteMem, 0, NULL);
    if (hThread == NULL)
    {
```

```
            _tprintf_s(
                _T("Could not create remote thread!\nError:\t%u\n"),
                    GetLastError());
            VirtualFreeEx(hProcess, lpRemoteMem, 0, MEM_RELEASE);
            return -1;
        }

        WaitForSingleObject(hThread, INFINITE);

        VirtualFreeEx(hProcess, lpRemoteMem, 0, MEM_RELEASE);
        CloseHandle(hThread);

        return 0;
    }

    int __stdcall RemoteFreeLibrary(HANDLE hProcess, HMODULE hLib)
    {
        LPTHREAD_START_ROUTINE lpLoadLibrary = (LPTHREAD_START_ROUTINE)
            GetProcAddress(GetModuleHandleA("Kernel32.dll"), "FreeLibrary");
        if (lpLoadLibrary == NULL)
        {
            return -1;
        }

        if (hProcess == NULL)
        {
            _tprintf_s(_T("Handle to Process 0x%x was NULL!\n"), (int)
                hProcess);
            return -1;
        }

        HANDLE hThread = CreateRemoteThread(hProcess, NULL, 0,
            lpLoadLibrary, hLib, 0, NULL);

        if (hThread == NULL)
        {
            _tprintf_s(
                _T("Could not create remote thread!\nError:\t%u\n"),
                    GetLastError());
            return -1;
        }

        WaitForSingleObject(hThread, INFINITE);

        CloseHandle(hThread);

        return 0;
    }

    int main(void)
    {
        char* szLibrary = PHYSICAL_PATH_TO_YOUR_DLL;

        HANDLE hMapping = CreateFileMapping((HANDLE)-1, NULL,
            PAGE_READWRITE, 0, sizeof(COMM_OBJECT), MAPPING_NAME);
        HANDLE hEvent = CreateEvent(NULL, TRUE, FALSE, EVENT_NAME);

        STARTUPINFO startInfo = { 0 };
        PROCESS_INFORMATION processInfo = { 0 };
```

```cpp
    BOOL bSuccess = CreateProcess(
        _T("C:\\Windows\\System32\\calc.exe"), NULL, NULL,
        NULL, FALSE, 0, NULL, NULL, &startInfo, &processInfo);

    if (!bSuccess)
    {
        _tprintf_s(_T("Error:\t%u\n"), GetLastError());
    }
    else
    {
        WaitForSingleObject(processInfo.hThread, 500);
        RemoteLoadLibrary(processInfo.hProcess, szLibrary);

        WaitForSingleObject(hEvent, INFINITE);

        PCOMM_OBJECT pCommObject = (PCOMM_OBJECT)
            MapViewOfFile(hMapping, FILE_MAP_READ, 0, 0, 0);

        if (pCommObject)
        {
            RemoteFreeLibrary(processInfo.hProcess,
                pCommObject->hLibrary);
        }
    }

    CloseHandle(hEvent);
    CloseHandle(hMapping);

#ifdef _DEBUG
    return system("pause");
#endif

    return 0;
}
```

6. 打开【解决方案资源管理器】,添加一个新的空 C++控制台应用程序项目 RemoteStartAddress。在【应用程序类型】中选择【DLL】,在【附加选项】中勾选【空项目】。右键单击【源文件】,添加一个新的源文件 main。打开 main.cpp。

7. 输入下面的代码:

```cpp
#include <Windows.h>
#include <tchar.h>
#include "..\common.h"

typedef struct
{
    DWORD dwProcessId;
    HWND hWnd;
} WINDOW_INFORMATION, *PWINDOW_INFORMATION;

BOOL IsMainWindow(HWND hWnd)
{
    return GetWindow(hWnd, GW_OWNER) == (HWND)0 &&
        IsWindowVisible(hWnd);
}
```

```
BOOL CALLBACK EnumWindowsCallback(HWND hWnd, LPARAM lParam)
{
    PWINDOW_INFORMATION pWindowInformation = (PWINDOW_INFORMATION)
        lParam;
    DWORD dwProcessId = 0;

    GetWindowThreadProcessId(hWnd, &dwProcessId);

    if (pWindowInformation->dwProcessId == dwProcessId &&
        IsMainWindow(hWnd))
    {
        pWindowInformation->hWnd = hWnd;
        return FALSE;
    }

    return TRUE;
}

HWND FindMainWindow()
{
    WINDOW_INFORMATION wndInfo = { GetCurrentProcessId(), 0 };

    EnumWindows(EnumWindowsCallback, (LPARAM)&wndInfo);

    return wndInfo.hWnd;
}

void ChangeWindowTitle()
{
    HWND hWnd = FindMainWindow();
    if (hWnd)
    {
        SetWindowText(hWnd,
            _T("Remotely started thread inside calculator!"));
    }
}

BOOL WINAPI DllMain(HINSTANCE hInstance, DWORD dwReason, LPVOID
    lpReserved)
{
    switch (dwReason)
    {
        case DLL_PROCESS_ATTACH:
        {
            HANDLE hMapping = OpenFileMapping(FILE_MAP_ALL_ACCESS,
                FALSE, MAPPING_NAME);
            if (hMapping)
            {
                PCOMM_OBJECT pCommObject = (PCOMM_OBJECT)
                    MapViewOfFile(hMapping, FILE_MAP_ALL_ACCESS, 0, 0, 0);
                if (pCommObject)
                {
                    pCommObject->hLibrary = hInstance;
                }

                CloseHandle(hMapping);
            }
```

```
                ChangeWindowTitle();
                HANDLE hEvent = OpenEvent(EVENT_ALL_ACCESS, FALSE,
                    EVENT_NAME);
                if (hEvent)
                {
                    SetEvent(hEvent);
                    CloseHandle(hEvent);
                }
                break;
            }
            case DLL_PROCESS_DETACH:
            {
                //
                break;
            }
        }
        return TRUE;
    }
```

## 示例分析

首先,我们创建了 common.h,定义了项目要使用的结构、事件和映射名。必须把程序分成两部分:exe 和 dll。为了运行应用程序,必须创建 exe;由于远程线程的开始地址在 dll 中,必须创建一个单独的 dll。在上一个示例中,作为应用程序本身,所有线程的开始地址都在同一个项目中,因为线程是在当前进程的地址空间中执行。而在本例中,线程将在另一个进程的地址空间中执行,这使得当前进程中的所有内存地址都用不上了。

当我们创建线程时,必须提供一个开始地址或开始执行线程的函数地址。如果在当前进程内定义,并使用函数的地址启动远程线程,那么调用 CreateRemoteThread 就会失败。因为当前进程中的地址属于进程私有,外部线程不能使用。因此,在 dll 内部定义函数和把 dll 加载进目标进程上下文很重要。我们可以使用 LoadLibrary API 加载 dll。

现在,我们分析一下 dll 本身。和应用程序(exe)类似,动态链接库必须有自己的入口。它的原型必须和下面的签名相匹配:

    BOOL WINAPI DllMain(HINSTANCE hInstance, DWORD dwReason, LPVOID lpReserved);

加载 dll 时,就从该方法开始执行。要区分各加载模式的原因,如下所示。

- ▶ DLL_PROCESS_ATTACH 方法通知:应用程序刚才加载了dll。
- ▶ DLL_PROCESS_DETACH 方法通知:应用程序刚才卸载了dll。
- ▶ DLL_THREAD_ATTACH 方法通知:加载dll的应用程序刚才创建了一个新线程。
- ▶ DLL_THREAD_DETACH 方法通知:线程存在,要执行一些清理工作。

我们在DLL_PROCESS_ATTACH中完成所有的工作。在主应用程序中,为了在应用程序和dll之间通信,必须创建一个文件映射。我们稍后解释。在动态链接库里，首先用OpenFileMapping API打开文件映射。如果打开了文件映射，我们把动态链接库的句柄写入hLibrary特征。我们稍后在结束工作时，要用到这个句柄从目标进程中卸载动态链接库。接着，调用ChangeWindowTitle方法。该函数尝试用FindMainWindow查找主窗口句柄。如果能找到主窗口的句柄，就能改变该窗口标题。

FindMainWindow方法使用EnumWindows API枚举所有顶层窗口。用户必须提供指向用户定义函数（即，回调函数）的指针作为EnumWindows的第1个参数。它的第2个参数是要被传入回调函数的用户定义对象，操作系统会为其枚举的每个窗口句柄调用该回调函数。我们可以利用这个特性用GetWindowThreadProcessId API询问属于线程（进程）的每个窗口句柄。

如果找到了一个属于目标进程的窗口，还要用IsMainWidow方法检查该窗口是否是主窗口。该方法调用GetWindow方法，第2个参数是GW_OWNER。如果返回的值是零，IsMainWidow方法将通过IsWindowsVisible API询问窗口是否可见，以确保如果窗口可见，它就是目标进程的主应用程序窗口。ChangeWindowTitle返回后，我们打开一个事件，该事件用于通知应用程序：目标进程中的工作已完成，而且可以卸载动态链接库了。我们用SetEvent API把该事件设置为signaled状态。

现在，我们来分析主应用程序。RemoteLoadLibrary方法用于在目标进程中创建远程线程。首先，要获得LoadLibrary API（LoadLibraryA——本例使用的多字节版本）的地址。该API将用作远程线程的开始地址，并带着指定的dll名作为一个参数来加载。以这样的方式，我们将把指定的dll加载至目标进程上下文中，并在里面执行。获取了LoadLibrary的地址后，必须在目标进程中调用VirtualAllocEx API分配足够的虚拟内存才能写入动态链接库的名称。注意，只指定动态链接库的名称是不对的，因为当前进程中的字符数组的地址在目标进程中用不成。

我们必须把动态链接库名写入目标进程，让字符数组的地址在目标进程中有效。分配好虚拟内存后，调用WriteProcessMemory API执行写操作。如果写操作成功，就用CreateRemoteThread API创建一个远程线程。必须提供有正确权限的目标进程句柄。不过，不用在句柄权限上太费心，因为应用程序创建了目标进程，其创建的返回句柄中就有我们所需的所有权限。

如果不能创建目标进程又必须打开一个现有进程，就指定 PROCESS_ALL_ACCESS 标志作为 OpenProcess API 的第 1 个参数。只有以这种权限返回的句柄 CreateRemoteThread API 才能使用。成功创建远程线程后，我们使用 WaitForSingleObject API 等待线程完成任务并返回。RemoteFreeLibrary 方法用于卸载 dll。首先，为了启动远程线程，必须获取 FreeLibrary API 的地址，该远程线程的任务是运行 FreeLibrary（参数是传入的 dll 的句柄）。以这种方式，远程线程将根据传入的句柄卸载指定的 dll。

在应用程序的入口中，首先必须把物理路径设置为已编译的 dll。这非常重要，因为目标进程中的远程线程将根据给出的路径加载 dll。该路径必须能绝对成功地加载。接下来，我们创建了一个文件映射，用于与 dll 通信。这也很重要，因为在卸载 dll 时，必须把它的句柄提供给 `FreeLibrary` API。只有成功加载后，目标进程中才知道被加载 dll 的句柄。因此，我们只能在与 dll 相关联后才能从目标进程写入文件映射。接着，我们创建了一个事件，将用在 dll 中触发主应用程序：句柄已被写入文件映射，可以卸载 dll 了。然后，使用 `CreateProcess` API 启动计算器。如果创建了进程，就在 `WaitForSingleObject` API 中等待指定的 0.5 秒初始化。然后，尝试创建远程线程，并等待事件对象。当事件对象被设置为 `signaled` 状态时，就可以使用 `RemoteFreeLibrary` 卸载 dll，并关闭所有已使用的句柄。

### 更多…

路就在前方，走向何方、能走多远，由自己决定。本书用 C++语言编写示例代码，涵盖了与并发编程相关的大部分主题。当然，如果要在程序中使用并行，还要进行更深入地学习。要管理并行执行，必须具备全面的知识，如操作系统、编译器、面向对象编程、执行、运行时等，当然还有 C++语言（如果用这种语言进行开发的话）。我们在此抛砖引玉，愿本书成为读者构建编程大楼的基石。在本书中，我们多次提到实践是最好的老师。准备就绪，开始美妙的创造之旅吧！

# 附录 A

**本附录涵盖以下主题：**
- 安装 MySQL Connector/C
- 安装 WinDDK-Drive 开发套件
- 为驱动器编译创建 Visual Studio 项目
- 使用 DebugView 应用程序
- 为 OpenMP 编译设置 Visual Studio 项目

## A.1 安装 MySQL Connector/C

为了使用 MySQL Connector/C，要执行下面的操作。

1. 打开Web浏览器，导航至 http://dev.mysql.com/downloads/connector/c/，如图A.1所示。

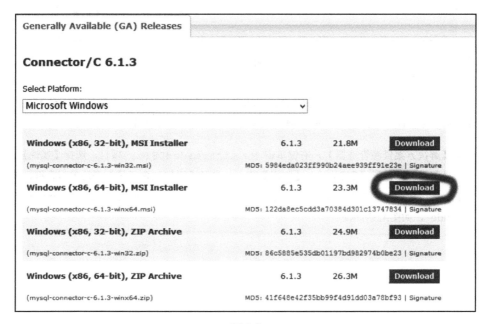

图 A.1

2. 根据计算机和项目的配置，下载mysql-connector-c-6.1.3-win32.msi 或者mysql-connector-c-6.1.3-winx64.msi 文件（仅供参考，以当前版本为准），然后安装，如图A.2所示。

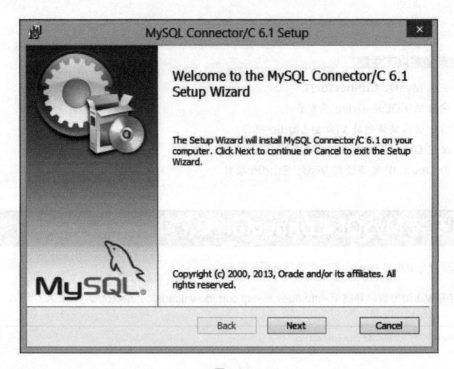

图A.2

3. 接受使用许可协议中的条款，选择【Complete】安装类型，安装完成后单击【Finish】。打开Visual Studio，然后打开 MultithreadedDBTest项目（第2.8节的示例）。

4. 打开【解决方案资源管理器】，右键单击 MultithreadedDBTest项目，选择【属性】。在【配置属性】下面，扩展【C/C++】-【常规】节点。打开【附加包含目录】旁边的组合框，单击【编辑】，使用Windows文件夹浏览，导航至 C:\Program Files\MySQL\MySQL Connector C 6.1\include ，如图A.3所示。

图 A.3

5. 单击【确定】。

6. 在【配置属性】下面，扩展【链接器】-【输入】节点。打开【附加依赖项】旁边的组合框，单击【编辑】，输入C:\Program Files\MySQL\MySQL Connector C 6.1\lib\libmysql.lib，如图A.4所示。

图 A.4

7. 使用Windows文件管理器，导航至C:\Program Files\MySQL\MySQL Connector C 6.1\lib。把 `libmysql.dll`文件复制到`C:\Windows\System32`中。

现在，就可以编译并运行 `MultithreadedDBTest` 项目或其他要使用 MySQL Connector/C 的项目了。

## A.2 安装 WinDDK-Driver 开发套件

为了使用 Windows Driver Kit，必须执行以下操作。

1. 打开Web浏览器，导航至 http://msdn.microsoft.com/en-us/library/ windows/hardware/hh852365.aspx。由于我们使用的是Visual Studio 2013，因此要下载 WDK 8.1 或者使用下面的地址：http://go.microsoft.com/fwlink/p/?LinkId=393659 。

   WDK 的下载页面如图 A.5 所示。

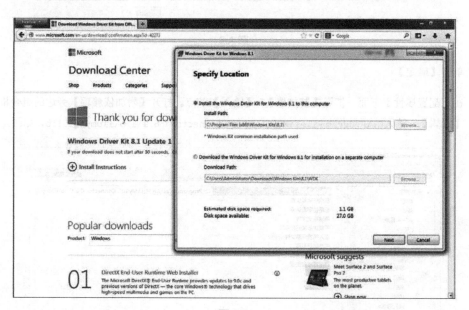

图 A.5

2. 单击【Next】，选择是否要参加**客户体验改进程序**，然后接受许可协议中的条款，开始安装。在Windows 中将看到如图A.6所示的画面：

A.2 安装 WinDDK-Driver 开发套件 295

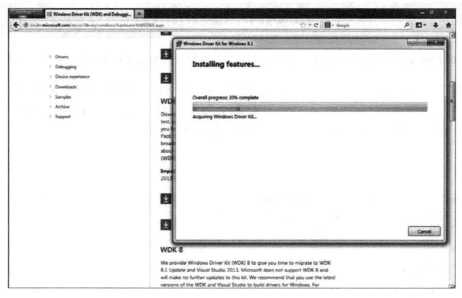

图 A.6

3. 等待安装完毕，关闭向导（见图A.7）。

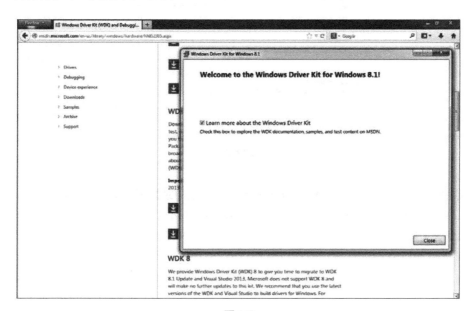

图 A.7

现在，成功安装了 Windows Driver Kit。

## A.3 设置驱动器编译的 Visual Studio 项目

为了使用 Windows Driver Kit 的特性（如编译器和库），必须设置 Visual Studio 项目属性，执行如下操作。

1. 如果已经按照第 2 章的说明进行了设置，打开 KernelThread 项目，在【解决方案资源管理器】中，右键单击 **DriverApp** 项目，选择【属性】。在【常规】下面，选择【目标文件扩展名】，选择【编辑】。把.exe 重命名为.sys。单击【应用】，如图 A.8 所示。

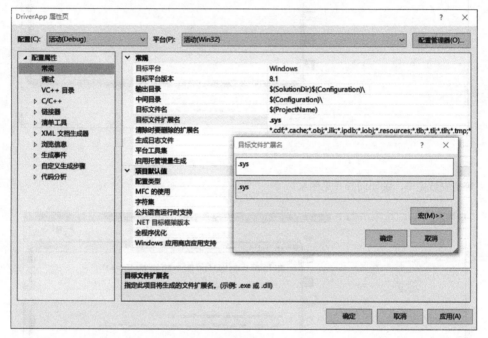

图 A.8

2. 现在扩展【C/C++】节点，在【常规】中选择【附加包含目录】。单击【编辑】，输入 Windows 路径：C:\Program Files (x86)\ Windows Kits\8.1\Include\km 。单击【确定】，然后单击【应用】（见图 A.9）。

A.3 设置驱动器编译的 Visual Studio 项目　297

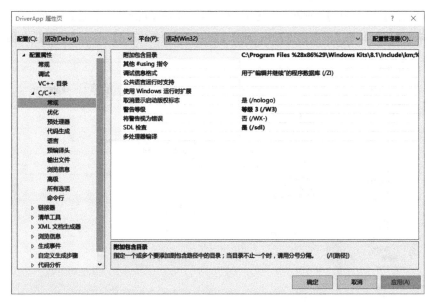

图 A.9

3. 现在，在【C/C++】下面选择【预处理器】，单击【预处理器定义】旁边的组合框，选择【编辑】。移除所有行，如果 Win32 配置激活，输入 _X86_；如果 x86 配置激活，输入 _AMD64_。如图 A.10 所示。

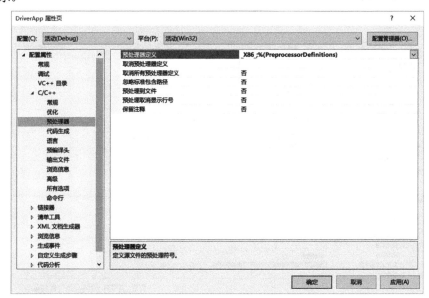

图 A.10

4. 单击【确定】并【应用】。现在,在【代码生成】中选择【启动 C++异常】,删除所有文本,使其为空,单击【应用】。在【安全检查】中,选择【禁用安全检查(/GS-)】,单击【应用】。如图 A.11 所示。

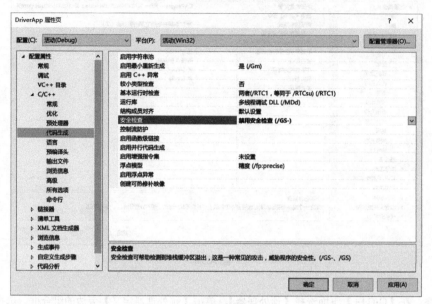

图 A.11

5. 现在,在【高级】中选择【调用约定】,并选择__stdcall(/Gz),如图 A.12 所示。

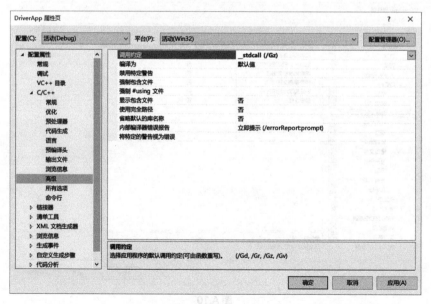

图 A.12

6. 现在，扩展【链接器】节点，在【常规】中，选择【启用增量链接】，选择【否（/INCREMENTAL:NO）】。现在，在【输入】中选择【附加依赖项】，单击【编辑】（见图 A.13）。输入 C:\Program Files (x86)\Windows Kits\8.1\Lib\win8\km\x64\ntoskrnl.lib 和 C:\Program Files (x86)\Windows Kits\8.1\Lib\win8\km\x64\hal.lib。

注意，根据激活的解决方案配置，把 **x64** 改为 **x86**。

图 A.13

7. 单击【确定】，选择【应用】。现在，单击【清单文件】，选择【生成清单】，在右侧的组合框中选择【否（/MANIFEST:NO）】，单击【应用】。现在，单击【系统】节点，选择【子系统】，在右侧的组合框中选择【本机（/SUBSYSTEM:NATIVE）】。选择【驱动程序】，选择【驱动程序（/Driver）】，单击【应用】。如图 A.14 所示。

图 A.14

8. 最后，选择【高级】节点，然后选择【入口点】。插入文本 DriverEntry。在【基址】后面，插入 0x10000。然后，选择【随机基址】，并删除所有文本，留白。然后，选择【数据执行保护】，删除所有文本，留白。选择【导入库】，并插入文本 ntoskrnl.lib。如图 A.15 所示。

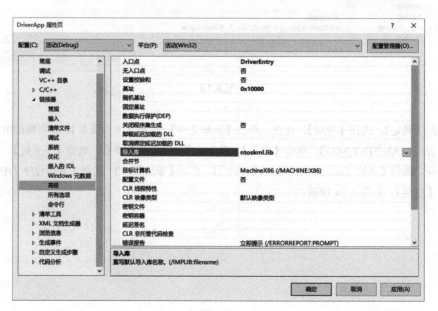

图 A.15

现在，可以编译并运行 DriverApp 项目了。注意，如果你使用 64 位操作系统，不可能开始数字无符号（*digitally-unsigned*）驱动程序，除非更改了**高级启动选项**（*Advanced Boot Option*）。Windows 的 Vista 和最新版本支持 F8 高级启动选项，也就是说，**禁用强制驱动程序签名**，不能为内核模式驱动程序加载时间签名环境，只为当前系统会话加载。该设置不存在于跨系统重启（摘自 MSDN）。欲了解更多信息，请访问 http://msdn.microsoft.com/en-us/library/windows/hardware/ff547565(v=vs.85).aspx。

## A.4　使用 DebugView 应用程序

为了从内核驱动程序中显示输出，要安装 DebugView 应用程序。执行下面的操作。

1. 打开Web浏览器，导航至 http://technet.microsoft.com/en-us/sysinternals/bb896647.aspx。下载DebugView ZIP文件，并将其提取为一个文件夹。运行DebugView应用程序，在菜单中选择【Capture】，勾选【Capture Kernel】或者按下Ctrl+K。如图A.16所示。

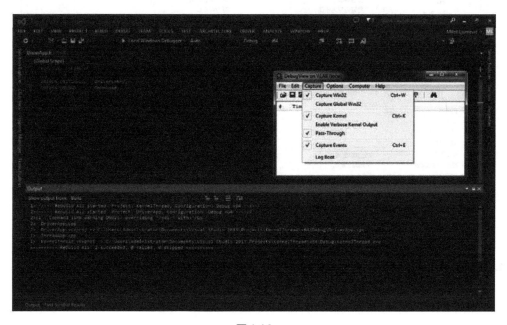

图 A.16

2. 现在，如果运行 KernelDriver项目，就能使用DebugView看到驱动程序的输出。

## A.5 设置 OpenMP 编译的 Visual Studio 项目

为了使用 OpenMP，必须要执行下面的步骤。

1. 如果按照第 7 章中步骤编写了 ConcurrentDesign 项目，打开该项目。在【解决方案资源管理器】中，右键单击 ConcurrentDesign 项目，并选择【属性】。在【配置属性】下面，扩展【C/C++】节点，选择【常规】，然后选择【多处理器编译】，选择【是 (/MP)】，单击【应用】。如图 A.17 所示。

图 A.17

2. 现在，在【代码生成】中选择【启用并行代码生成】，并选择【是（/Qpar）】，单击【应用】。如图 A.18 所示。

A.5 设置 OpenMP 编译的 Visual Studio 项目    303

图 A.18

3. 现在，在【语言】中选择【OpenMP 支持】，并选择【是（/openmp）】，单击【应用】。如图 A.19 所示。

图 A.19